Normally, guys will take girl numbers date. Pick her up, take he[r] while, take her home, have that aw[kward] kiss, and maybe be invited inside. [She's in] control of what'll happen later. Th[at's not] a date like that, so I change it up a bit.

What I do is ask her to come pick *me* up, and when she comes over, I answer the door nude except for a towel.

"Oh, hey, sorry, I'm not ready yet," I say. "But come in." Then I stop and make a face, the kind of face girls tend to make when they ask if you have a condom, and I ask "Would you mind taking off your shoes?" I gesture with a tilt of my head to the shoes sitting there by the door, and she always hesitates, then smiles uncertainly and slips her shoes off, looking at my washboard abs and the towel wrapped around me, and she comes into my territory feeling a little vulnerable and at an advantage at the same time.

"I'll just be a minute," I say. "You want a glass of wine while you wait?"

She makes a yes-no tilting of her head and says "Sure."

I lead her into the kitchen in my towel, her feeling the cold tiles underneath her bare feet and feeling at home in a stranger's house, feeling the tingle of being with me in such a familiar and personal way before our first date even begins. We toast, watch each other over the rims of our glasses while we drink, and then she looks around the house and invariably asks me about one of my paintings. We end up in the room I use for my studio, which is right next to the bedroom. She is always amazed and never fails to give me a sly smile, biting her lower lip and telling me she wants me to draw her.

"Really?" I ask.

She nods. We never end up making it to dinner.

Also by the Author

The Heresy Series
By The Sword
Friends Like These
The Kettle Black
After The Flesh
The Camel's Back

The Icarus Trilogy
In Shining Armor
A Mind Diseased
Thy Neighbor's Wife

Comedy Adventure
Saga of the Beverage Men
The Prince of Foxes
Navarre
A Brothel on Hibiscus Lane
Val & el Templo de Chocolate

Memoirs of a Swine
Just Plain Trouble
This Little Piggy
A Bad Husband
No Good Deed
An Empty fist

Nonfiction
Variety Is the Spice
A Thousand Words
The Hero Mindset

MEMOIRS OF A SWINE

Saint Lucy's Eyes

For Patrick —
It was a pleasure to meet you, and I hope you enjoy the book! — Alexander Ferrar

Alexander Ferrar

NOTE: If you purchased this book without a cover you should be aware that this book is stolen property. It was reported as "unsold and destroyed" to the publisher and neither the author nor the publisher has received any payment for this "stripped book." Kind of like that free music you downloaded. Prick.

This is a work of fiction. All of the characters portrayed in this book are fictitious, and any resemblance to real people or events is purely coincidental. Should any resemblance seem apparent between the characters or events portrayed in this book and you or your own life, then the word "Memoir" in the title oughta be a big, glaring clue. Yeah, paybacks're a bitch, huh?

MEMOIRS OF A SWINE

Copyright © 2011 by Alexander Ferrar

All rights reserved, biatch, including the right to reproduce this book, or portions thereof, yadda yadda yadda, in any form. And we ain't kidding, neither.

ISBN: 1514627574

Published by Bunbury
First edition: July 2015
Second edition: December 2015
Third edition: May 2016
Fourth edition: September 2018
Fifth edition: March 2021

Cover photo by conrado on Shutterstock, design by Alexander Ferrar

A Note on the Text

The text of this novel was set in Charlmesby Oldface, a font devised by the obscure Flemish typesetter Farleswick Charlmesby, who designed it as a passive-aggressive way to make his wife leave him. Many men will choose this method, acting inappropriately until their very presence in the house becomes intolerable and their partners, fed up, then take the initiative to end the relationship, thus taking the responsibility of confrontation out of his hands and absolving him, at least in his own mind. Anyway, it worked.

www.alexferrar.com

This novel is dedicated to all of the hearts I broke before I realized what a jerk I'd been.

Introduction

There were three reasons for writing this book. One was as a rebuttal to pick-up artist handbooks. I have read a lot of really bad advice in them and feel someone ought to set the record straight. First off, that palmistry crap is for schmucks. Pick-up lines, too. Forget 'em.

And some of that stuff the so-called masters are saying, like going to the ATMs in the better side of town and rooting through the garbage next to them, where people throw their receipts? They say when you find a receipt for an impressively-large withdrawal, you keep it handy in your pocket for when you want to give your number to some girl you just met. You write your number on the back, and she is guaranteed to call. But that's pretending you're something you're not. What's the point? She's going to find out, sooner or later.

Ditto walking into a crowded bar alone, and pretending to point and wave at people you don't know, so you'll look popular to anyone who happens to be watching. That's not how to be a man. That's how to be a poseur.

Another reason for writing this book is the astonishing number of books on how to be a man. That is something I've taken an interest in, and I find the advice people are spouting to be weak. We don't need to know how to do a convincing comb-over. We need to know how to be bald and proud of it. We don't need to know how to wear jewelry tastefully or how to pronounce French wines without sounding like a poof. We don't need to be okay with who we are. We need to know how to be better.

We *especially* don't need to know how to go running to a teacher and tell on a bully. We need to know when to hit first and where, and how much is enough. And how to not be a chicken when someone's holding a gun to your head.

So, I wrote this, based on how I learned. The hard way.

I wrote it as a novel, and some of the names I changed to protect the guilty. It turned out to be rather long, so we're breaking it up into a trilogy.

About the name Johnny Yen, it was an alias I had for a while, like Rabbit (some black dudes used to call me that, and it stuck). My middle name is John, after my Grandfather and my uncle, and there's a Johnny Yen in two songs, one about an artist who's always doing over-the-top crap to get attention, and another, *Lust for Life* by Iggy Pop. Both seemed fitting.

And now the third reason, because it is a true story. It is based on a series of crazy adventures I had, during my salad days, when I was green in judgment, where I both roosted in henhouses and wallowed in pigsties, and writing this is my way of making lemonade.

We expats all like to swap stories of how we came to be in Antigua, the same way born-again Christians love to tell how they came to know Jesus. How we found our way here. This is probably true of expats in every country, but I can only speak for us. Oftentimes, the stories are hilarious, and I make vague plans to compile a book of them.

For example, one friend was a drug dealer with a hot crazy girlfriend. She asked him to loan her $1,000 for an investment, and he readily paid it. He left to go check his production line, and when he came home there were 1,000 scratch-off lottery tickets scattered on the floor. Big fight, lotta screaming, break-up, slamming door. Popped a beer and sat down, deep sigh.

Picked up a ticket and absently scratched it off. *Boom!* One free trip to Antigua, Guatemala. He came so he could distract himself from his broken heart, and while he was here, his little operation got raided and everyone but him went to jail.

Stories like that. Well, this trilogy (of four books) is mine.

How, before I became the Renaissance Man—owner-chef of a very popular restaurant, author of now seventeen books, and successful Baroque artist—I was just a hard-drinking. bar-fighting, heart-breaking prick. I hope you like it.

Prologue

In the marvellous city of Antigua, in the Land of Eternal Spring, there is an exotic ice cream shop across the street from a sports bar with jacked-up prices, and every day a man and a woman pretend not to see each other while they open them. It is the subject of much gossip between their neighbours, who wonder why the man, who smiles and says Hello to everybody he sees, would somehow not notice such a beautiful woman, and why she, who lights up every room she enters, finds herself fascinated by the cobblestones beneath her feet.

The waiters and bartender of the sports bar all know the ice cream man, and wave to him when they see him, and once or twice have noticed her flinch when they call his name. Since the ice cream shop has a lot more business, they'll sometimes stand at their door and watch people gush over whatever bizarre flavours he has that day, and remark upon the change that comes over him when the last customer in a shoal walks away. His smile fades, and his eyes glaze over, until he shakes his head as if to clear it. When he feels the burn of their eyes upon him, he glances over and does an upward tilt of the head that is somehow still a nod, and they'll wave again and go back inside.

One day, one of the waiters came across the street. He was a stocky Guatemalan with cinnamon skin and boy-band hair, and the ice cream man was ready with his smile.

"Hey, you're from the States, right?" the waiter asked in Spanish.

The ice cream man shook his head. "Bahamas."

"You're kidding? But I thought Bahamians were black."

The ice cream man laughed. "Well, the black ones are."

They beat around the bush for a while, talking about the flavours of ice cream, and how long the shop had been open, and which came first, the ice cream or the restaurant around the corner, and blah blah blah until the waiter was ready to ask "So, when're you going to come over and have a drink?"

The ice cream man made the face the waiter expected to see, and he was satisfied for the moment. *Slowly*, he thought.

The sports bar had no sign, and was only open after six on weekdays, because it was right next door to the Muni and they still hadn't gotten their paperwork done. The ice cream man wondered if they thought the clerks of the Muni only lived to work and went home to sleep in caves as soon as their shift was over, living in ignorance of what went on in their city for the rest of the day. It had opened a month before, and was a sister business to Bistrot Cinq, down the road, the same way the ice cream shop was a satellite of the restaurant art gallery around the corner.

The beautiful woman had been the manager of Cinq for only three months, after losing her job as chef at a golf resort several miles outside of town. Rumours abounded as to why, but no one knew for certain. So, she'd come back to Antigua, and when the owner of Cinq offered her the second job managing the new place a few hours a day, she was hesitant, knowing she would have to walk past either of the Bahamian's two businesses to get there.

"So, Bahamas," the waiter said. "Why would you want to leave *there?*"

"Oh, I left when I was eight. My parents moved to Florida, and then we went to live in Ireland for many years. Then I came here."

"What brought you here?"

"Long story."

The waiter shrugged. *"Es silencio."* Meaning, there are no customers, so what else is there to do?

The ice cream man hesitated, wanting to get something off his chest and thinking he'd rather trust a complete stranger than a priest. He looked at the nosy waiter for a long moment, wondering how much he was going to tell him.

I

> "I think people often try to find through sex
> things that are much easier to find in other ways."
> —Patricia Highsmith

Okay, first off, lemme say this: nobody can tell you how many times you can be in love or how often or after how long. I fell madly for two different women on one day in Taormina, Sicily (for real, they were the *one,* both of them) and then again with a hot dark waitress in Venice a week later, and anyone who scoffs just doesn't know how to live. Now, that trip to Italy was years ago, and since then I've been in love maybe only three or four times, tops.

This time, though, it was different.

But back up a minute.

I'd been reading up on what they (apparently) call the Venusian Arts—not because I needed to, no way, definitely not, but maybe just to beef up my game a little, because it never hurts. The first thing I learned was that Less is More. Canned pick-up lines and slick dance moves and flashing your money are nothing compared to just giving her a look. Not a look like "I want you" or "Wow, you're gorgeous" or "You'll have the night of your life with me," and not a leer or a stare or an appraisal. Just making eye contact until she looks away, then looking away yourself until you feel her eyes on you, and meeting her gaze a second time. Yunno in Madonna's "Papa Don't Preach" video, where she's hanging with her girlfriends, wearing that Italians-Do-It-Better t-shirt, and she sees the guy her father warned her all about (the one he said she could do without) and their eyes meet? *That* look. It's magic.

So, I'm in Rome this past summer with my Dad and two brothers and two of our cousins and our grandma, and I can't get that look right to save my life. I'd done it a bunch of times in the past by accident and it worked just fine, but I couldn't contrive it worth a damn. Either I was coming off as sly or dismissive or creepy...not like I saw anyone I really wanted to pull, but it's practice for when I do, yunno?

Whatever. Anyway, skip ahead, skip ahead, skip ahead.

We get to Corsica and both my brothers are too hungover to do anything but lie in their beds and grunt, so just my cousins go with me to the top of Bonifacio and take pictures. We get a bunch of landscapes and some of this elaborate marble tomb in the graveyard, and some of us just posing, when I try just a plain ol' glance at this one chick standing with her boyfriend as we're passing by. We lock eyes.

Now, y'ever get spotted by a dude when you're eyeing his chick, and he decides to make out with her? Kind of a passive aggressive non-confrontational marking of territory? Used to be, the guy'd come up on me swinging, but since I've been going to the gym and drinking shakes and popping pills they aren't so quick to throw down, so, instead, they grab the girl and kiss her. And show tongue, as if I wouldn't get the hint otherwise. And I'm watching this thinking "Take that, dead horse! And that! And *that!*"

So, that's what this guy does, this guy with long, shaggy black hair, and I'm startled to see her protest. She's resisting, saying something in a language that I don't recognize, but the meaning is *all* too clear. And she's looking at me while she's saying it, with her gray-blue-green eyes that are all I can see as I watch over my shoulder. He's saying something back, insisting, but she's having none of it.

We leave the top of the mountain and make our way back to the city, and I try that glance on a few other women that we pass, but they're too into their boyfriends or husbands to respond.

Now, I feel I ought to add that I'm not like every other guy you'll meet. I may have only slept with forty-something women, but already I'm (for lack of a better word) bored with it. It's the *chase* I enjoy, the anticipation, and when I get it right away, I'm disappointed, even a little depressed the next day. When I meet some girl at a club and kiss her all the way to my car, if that isn't as far as we go, in less than an hour I'll be wondering with each thrust if this is what it's all about. If *this* is what's so coveted by all the world, the end all be all. 'Cause if it is, it is *so* overrated.

That's why I'm into creating sexual tension and longing and frustration, to build it up, and sometimes I'll just let the whole thing drop because I know I'll never get disappointed that way. Anyone can tell you that the sex you imagine having is almost always better than real life. She'll always remember you doing such-and-such, and no matter how good it was she'll eventually get tired of reliving it in her mind. But if it *doesn't* happen, she'll spend the rest of her life hypothesizing what it could've been like. It's a kind of immortality, and I can live with that.

So, I'm reflecting on this while we're in a souvenir shop, looking at the kind of useless stuff that only reminds you later of having been in a souvenir shop. "I went all the way to Corsica to buy this coffee mug/tee shirt/whatever." Stuff you can buy anywhere with a different name on it. Moving out to the front of the store where the postcards are, I see the shaggy black hair and oversized shades of the boyfriend passing in front of me, and right behind, a pretty face with chin-length brown hair that turns gray-blue-green eyes straight into mine. They widen a little, and her face jerks back to stare straight ahead, and I come out of the shop to watch them go.

She's wearing a short black dress and flip flops, and she is a little big-boned, but not in a bad way. A sturdy kind of way, but with very little fat. Healthy. Not like those skinny girls that're so popular nowadays that I hate, because making love to them is like humping a bag of chisels. The boyfriend, he's wearing flip flops too, only her feet're clean and his aren't, and his maroon tee shirt says California something.

Just as I'm wondering where they're from, she glances back to see if I'm looking at her still and this confusion flashes across her face, like "What's with this guy? Why's he *looking* at me?" so I turn my gaze to postcards, pulling my shades out of where they hang in the point of my v-neck, and putting them on so she can't tell where my eyes are fixed. When I see her again, she's drifting away from Maroon California and looking at me uncertainly.

I let her look at me for a moment, because for a tourist I'm close enough to handsome, and I'm dressed pretty well. My eyes cut sideways to my reflection in the souvenir shop's glass door, propped open with a ship-in-a-bottle, to check my hair's not too windblown, and I'm alarmed that it is. I cut my eyes back and get my first really good look at her face. She has Cupid's bow lips that are pursed in concentration and a strong jaw with a pointed chin. Above all, she has the face of a sweet person, which I've always preferred.

I turn my face back toward hers full-on and she snaps her head away, suddenly absorbed in all of the different ice cream flavors on display outside the gelateria. I cross the cobbled street to a café, order an espresso so I won't feel guilty about using their bathroom, and go straight for the sink while they're preparing it. A little water to wake up the gel in my hair from last night, and I'm spruced up enough to look like I didn't do it on purpose. I shut my eyes and take a deep breath, listen to the hum of the air conditioning and the ringing in my ears, feel the weight of my body and relax.

Outside, I down my espresso and go into the street.

I can't believe it, but I feel like a teenager again, with my heart shuddering and my senses heightened, and the paranoid fear that all my thoughts are on the air.

She's not anywhere in sight, so I go back into the souvenir shop to look at Corsican shot glasses and ashtrays with my cousins. Didn't want to be seen too much—that's the second thing I learned about attraction: if you want to feed it, you've got to keep it hungry. Eric and Jorge, my cousins, they say we should go across the street to the outdoor café and have a little hair of the dog, and I couldn't agree more.

We debated the merits of one beer over the other, passing around our pints the same way we all shared our food at the restaurants every night. The two of them had all kinds of criticisms for the three different brands, but I confess they all tasted the same to me. Like Coke and Pepsi. Sorry. And while we're all catching up on what's different in our lives that we can't talk about with my Dad and our grandma around, I've got one eye on all of the girls passing by. Jorge tells me about how his son is going to Pre-K and how well he's doing, and how weird the mother is getting, and we finish our beers and get up to go when I spot Maroon California out of the corner of my eye.

The two of them, they're walking past us holding hands, both with their sunglasses on, and I can't tell if he's bristling because he recognizes me or he always walks that way, but I don't wonder long because she slows up ever so slightly to hang behind him, and takes off her shades. With a quick glance at him to make sure he won't see, she turns her head and gives me a full-on stare. In that fleeting moment that lasted an eternity I felt my face being soaked up like...like what? How do you describe the feeling of someone memorizing your image? In those bright gray-blue-green eyes I saw "I want you" and "Wow, you're gorgeous" and "You'll have the night of your life with me" and I knew right then that the next time she slept with Maroon California she'd be pretending it was me, and she'd tell all her girlfriends back home in Whereveria, and because it'll be her imagination instead of me making love to her, I'll be the best lover she'll ever remember. She tells me all this with those eyes that I find myself lost in, and in a flash, *whoosh,* she's gone.

I dunno how long I stood there, stunned, watching the corner she'd disappeared behind before Eric gave me a shove.

"Ya might wanna close your mouth," he said. "Flies're getting in."

I heard someone holler my name, and Jorge pointed to my brothers coming through the milling tourists. We all sat back down at the table and ordered more drinks. The four of them yakked away about God-knows-what, but I was too busy staring at that corner to pay attention. Thinking, were those two just boyfriend-slash-girlfriend, were they on their honeymoon, what? I don't imagine you take some chick you're just dating to Corsica, excepting maybe you live in Sardinia or something. How long had they been together? What were they planning? Or were they just playing it by ear? Only thing I knew for sure was, maybe this morning when they woke up next to each other she loved him truly with her whole entire heart, but she certainly didn't now.

Did he pay for this romantic trip out here and it's here that her love for him ended, just like that, over a perfect stranger? Third thing I learned was, life's like that. Attraction is not a choice. It'll come on all over you even despite prejudice and there's nothing you can do about it, or just as quickly it can vanish. You can't decide to want someone—either you do or you don't, and the second you stop it's over for good, and they can't reason with you, persuade, cajole, coerce, or anything. Can't do anything but drive you away further.

And it doesn't matter who someone is, what they look like, or whether you've ever even heard the sound of their voice, much less been fascinated by something they've said. Sometimes it just happens, *bam!* And you can't get them out of your head, those gray-blue-green eyes staring into yours like a Cheshire Cat grin long after the rest of the face has gone.

I spent the next hour or so searching all of the faces I passed until I gave up, and resigned myself to looking at driftwood dream catchers and jewellery made out of indigenous shells, t-shirts and coffee mugs and postcards. Odd, I thought, that my little exercise had backfired, but somehow strangely wonderful. All I cared about was seeing her again, just to have a moment alone together and be lost in her. I felt a tingling drunkenness like I remember feeling after my first kiss, and some of the ones after that.

We turned into some other store and I was shocked to be

bumping into a maroon shirt and looking up into the unshaven face of Whatshisname, the shaggy-haired guy, the boyfriend. I said "Pardón" and he gave me a curt little nod, brushing past me, and there she was. Not three feet away. My heart seized up inside my chest, my throat closed. Her profile was close enough to touch as she was looking at leather handbags and something-or-others and I wished she wouldn't notice just long enough for me to memorize every detail of her face. But she glanced up and saw me, and her Cupid's bow lips spread wide in a warm smile as she stepped past me to follow her boyfriend. I realized I had bitten my lower lip, but didn't let go of it until I'd turned around and seen him take her hand and lead her away, and she peered back over her shoulder the way I'd done when we first locked eyes, smiling broadly at me as she walked out of my life forever.

I didn't tell anyone else about what'd happened, not because they'd tease me for not pursuing, but because I thought it would somehow cheapen it. That night we ate dinner at the outdoor pavilion of a nice restaurant near our marina, and it might've been a coincidence, but the veal chops in sour apricot sauce was the best thing I'd ever eaten in my life, the wine that my family complained about was wonderful, and when a waiter knocked over a glass and spilled my brother's water into my lap, I couldn't bring myself to care.

I spent the entire evening keeping a vigil, watching the sidewalks on either side of the pavilion, checking every face that came into view. Desperate to see her just one more time, even though I knew it would be better if I didn't. I knew I'd step out of my chair and weave my way in between the tables to stop her, take her face in my hands and kiss her deeply.

No matter *where* the boyfriend was.

I bought a bunch of souvenirs the next day. Corsica has got some indigenous shell they call "Saint Lucy's Eye" that's got a cute little swirl in it, and they exploit the hell out of it in earrings, necklaces, bracelets, and whatevers. I bought one of each, ostensibly to give to my girlfriend when I went home. I also had to buy a Corsica shot glass, coffee mug, espresso set, ashtray—anything that would remind me of this great town. City, whatever.

I vaguely remember from Sunday School that Saint Lucy is the one you prayed to if you'd gone blind or needed glasses or something, because she could restore your sight. It seemed appropriate somehow but I try not to think about it too much.

We're in Sardinia now and I've seen no women that interest me at all. I wonder if I'll feel the same way when I get home. I wonder if I'll pretend the girl I'm seeing is this one with the chin-length brown hair and the gray-blue-green eyes and the Cupid's bow lips. I don't even know that girl's name, or what country she's from, or how to contact her or what I'd say if I did, but I have a funny feeling I'll never forget her. Like she's achieved a kind of immortality.

And I can live with that.

II

The same way amputees can still feel a phantom limb and the recently dumped feel the lingering presence of their ex-lovers, I always feel my goatee for a few days after I've shaved. I trace the lines where it used to be with my fingertips and try to tug on my moustache, and I feel stupid. Lord knows how dumb I look doing it.

That's what I was doing at one of those outside tables at Bellagio, in City Place, when my phone rang, the grandmother of the girl of my dreams on the other end. Or so I'd delude myself into thinking later that day.

"I need your help," she rasped at me, her voice sounding like the crackle of dead leaves crunching underfoot. I couldn't help but imagine the tiny white hairs quivering on the lips of her bitter, pinched, evil mouth while she spoke. "I'm throwing a small cocktail party tonight as an excuse for you to come over and do whatever you have to do—seduce, I don't care—but *get* my granddaughter away from her *jackass* boyfriend. I mean, wait'll you see this guy's *shoes*. They're awful. He came to Palm Beach *knowing* he couldn't set foot in anywhere worth going without at least loafers and a blazer. Christ, he didn't even bring a *blazer!*"

"The infidel," I said, grinning.

"Beatnik, if you ask me. So, be here at seven? Bring your mother so it'll look legitimate."

"I'm sorry, Mrs. Morrow. I already have plans."

"She's gorgeous, Johnny. You'll like her."

"I'm sure I will, but—"

"And I want you to paint her portrait."

"...see you at seven-thirty, Agnes."

"You're a good boy, John. See you then." Click.

Good boy? Hmmmm. I just accepted a bribe from an old lady pimping out her flesh and blood. She knew I hadn't sold any paintings since Christmas, three months earlier, and would love to have a project and some money to spend. The paint wasn't doing any good staying inside the tubes, and money-wise, I was knee-deep in red ink.

Now, back up a minute.

Mrs. Morrow and I went way back to maybe the mid-90s when I was in my teens, and her visiting grandson needed a babysitter while she and the parents went out to some gala or other. The kid said he had so much fun without doing anything bad that, whenever any of her many granddaughters were in similar situations over the next few years—the ones of a riper age, at least—she'd ask me to take them out and keep them from getting bored. I'd always groan, expecting it to be some pity date, the same way the lucky girl would expect me to be some "nice boy," and we'd both be pleasantly surprised when Agnes opened the front door and we future lovebirds saw each other for the first time.

Mrs Morrow had many beautiful granddaughters, and but for the grace of God she almost had a few great-grandchildren before it woulda been convenient.

Anyway, it was easy to cancel my plans for the evening, because they consisted of merely going to the bar where Josh and them worked, and adding to my already unwieldy credit card debt. Maybe sleeping with one or two of the chicks there, if there were any worth looking at.

It's safe to say that, in the seven or eight months since my magical trip to the Mediterranean last year, I'd cut eleven more notches in my belt, bringing me up to fifty-something, and not one of them had gotten to me the way that girl in Corsica did.

And, mother of all tragedies, I had gotten bored with it. There was no more sexual tension, just sex sex sex. No more thrill of the hunt, no more elation, no more satisfaction of conquest. Made me wonder sometimes why I even bother.

I finished my coffee and signalled the waiter, some young chico posing as an Italian fresh off the boat—part of Bellagio's aesthetic—speaking basic Italian phrases with a strong Cuban accent. He nodded and brought the check over, and I turned to my mother. She sat across from me, chain-smoking and staring off into space.

"Thanks for lunch, Mom."

She nodded absently.

Elizabeth Morrow looked exactly the way you'd expect a girl of that name to look; skinny, posh, stylishly dressed, comporting herself well and using words like "comport." While she may not've been especially pretty, she was attractive, with dark red, almost black Aeon Flux hair that somehow did not look cyber-punk in Agnes's salon. It looked stylish and sophisticated. Her non-geographical cream-of-society accent didn't sound affected; it sounded cultured and refined. And her dress didn't look purchased with a stolen credit card; it looked like a gift from a doting uncle or a sugar daddy.

And—this might come as a shock, but—we were both pleasantly surprised when Agnes opened the door and we future lovebirds saw each other for the first time.

"John! Connie! So glad you could make it, come in, come in," Agnes said, gesturing into the house with her eagle's talon of a hand. Adding to her bird motif, her spectacles had gotten even bigger, in the current sunglasses fashion, until looking at her made me think of an owl. Her vowels came out in a long drawl that smelled like Grey Goose on the rocks with a splash, and her huge eyes narrowed conspiratorially as she leaned into us and stage whispered "Just wait'll you meet him! Wait'll you see his awful *shoes*."

I smiled, pretending not to mind the flecks of spit on my chin from her malevolently hissed S's and said "Thank you for having us," kissing her on the cheek. My mother exchanged glowing compliments and hugs with her as we were escorted in. My eyes were surreptitiously scanning the room as I smiled at the widows, nodding at the ones I knew, and I had just opened my mouth to start greeting them when I met Elizabeth's gaze.

She was holding a tray of hors d'oeuvres, frozen, stooping

to place it on the Louis Quatorze occasional table in front of two matronly older ladies on the Louis Quinze swan fainting couch, arrested in the middle of the action. Her hair was in front of her face, her lips parted in surprise, and when she remembered herself she straightened with the tray still in her hands, regaining her composure and mustering some drama class dignity.

I made my way around the sofas and armchairs, shaking hands and kissing cheeks, trying to decide whether to negate her by meeting her in between all these others or saving her for last. Negating a woman was either the sixth or seventh thing I had learned—showing her I really didn't care whether she liked me or not, but not in a dismissive or arrogant way. Just showing she wasn't *that* important. Not easy to do.

While I was saying Hi to Mrs. Koontz and hearing about how small and adorable I was the last time she'd seen me, and what a fine man I'd grown into, I was startled to see Elizabeth step right up to my left side and offer her hand, interrupting me as I started to feign modesty.

"You must be John. It's a pleasure; I've heard a lot about you."

"Pleasure's mine," I said, making an effort to affect disinterest. "Didn't get your name?"

"Morrow," she said. "Elizabeth Morrow."

Like Bond, James Bond.

And, the damnedest thing, I wasn't in the least bit charming for the next few hours. We made small talk, all of us, until we had had a little too much to drink and started talking about important things like art, philosophy, and culture. The sudden rise in volume awakened the boyfriend from his widdle nap and he finally joined us. In just a shirt, untucked, the top two buttons undone, and khakis. And olive green gingham canvas slip-ons with designer holes torn in them. And white socks.

At a posh cocktail party on the Island, for Christ's sake. Even I had to frown on that. He came in and flopped carelessly down into a Louis Quinze bergère with foliate-carved golden beechwood and blue-beige upholstery, the way I tend to do whenever I'm not at posh cocktail parties with society women, and sat there as one does when he *has* to be there without wanting to be. He hadn't shaved, either.

"So, you're all here."

"Everybody, this is Charlie," Agnes said, her voice just sickeningly sweet, which everybody but Charlie understood perfectly well.

"Interesting shoes," I said to him, pointing out the elephant in the room. He extended one leg, the leg he had hooked over the arm of the chair, and pivoted his ankle so he could look at his shoe and consider it, as if he hadn't before.

"Yeah, I like 'em. Cost me a fortune, though."

"I'll bet. The place you bought them, do they sell grown-up shoes there, too?"

I felt the air in the salon get displaced ever so slightly by all of the eyes widening, felt the pressure change from all the sudden intakes of breath.

Charlie the Gauche didn't even notice.

"Nah, just cool shit. I got this belt there too…oh wait, I forgot to put it on. Oh well. Another time."

"Would you like a drink?" Elizabeth asked him, pointing toward a Louis Seize parquet game table with the bottles and glasses and ice bucket. "There's some—"

"Whatever's fine."

Demonstrating some good breeding, she took the hint and got up to fix him a drink without calling attention to his rudeness, keeping the collective mood light. We went on talking about the zeitgeist until the guests of more advanced age excused themselves, calling on their drivers or companions to assist them. I stood up and helped a few unsteady women to their feet, and scored points with all the guests, but only *partly* for that reason, mind you. For the record, I mostly did it because it was the thing to do.

More and more of them excused themselves until it was just Agnes, my Mother and I, and Elizabeth and Charlie the

Gauche. Elizabeth changed from her high heels into flip flops and, when she caught me checking out her well-shaped feet, locked eyes with me. For a second, I saw that look again, the one that had me so stunned seven or eight months earlier.

"Who wants to join me outside for a cigarette?" she asked the room, without taking her eyes off of mine.

"Oh, I will!" my mother piped up.

"Yeah, but mine're in the room," Charlie said, meaning "Yeah, I'll smoke one of yours."

Elizabeth, still looking at me, managed to roll her eyes without actually rolling her eyes. I fished my pack out of my jacket pocket and said to Charlie "Here, want one of mine?"

"Oh! All right, thanks, man," he said happily, wiping his hands on his trousers before reaching into the pack for one. He didn't even notice the air crackling between me and his girlfriend. Hmmm, I thought. Curiouser and curiouser.

We went outside, all of us except for Agnes, and Elizabeth spent a little longer than she needed to steady my hand with her own as I lit her cigarette. The way she smoked, she kept her forearm vertical with her elbow pulled into her side, the hand limp with her palm upwards.

"Soooo," she said to me, and paused long enough for me to know this would be a loaded question. "My grandmother said you're going to take us out and show us a good time. What do you have in mind?"

I shrugged, using only my lips and eyebrows. "Not much to do here on a Wednesday night, 'cept this bar I know."

"I'm up for anything. We leave tomorrow and I've done *nothing* for six days. We. We've done nothing."

"That's not true," Charlie said. "Grandma took us out—"

"Please don't call her Grandma, okay? Didn't you see that look she gave you when you called her that the other day?"

"She'd just bit into an olive. I figured she didn't like olives."

"She's almost ninety years old, Charles. If she hated olives, she would have known it by now and not have them in her house."

"Maybe she keeps them around for guests."

"She likes olives, Charlie. Jesus!"

I thought she was going a little overboard showing me she wasn't happy with this clown. Take *that*, dead horse! So I changed the subject, a blind stab at something I had overheard her talking about earlier.

"So where's this spa I hear you own?"

She brightened. "It's a day spa in Antigua, doing very well not just with the expatriates living there, but also the hundred million backpackers that're coming through pretty much all year round. Most of them are taking a break from life after college and before buckling down to work and becoming adults. Or, they're fugitives. You wouldn't believe how many people there are on the run from something or hiding out, and always having to look over your shoulder can be pretty stressful. Nothing like a full body massage to alleviate the stress of being on the lam."

"Is that your ad campaign?" I said, laughing.

"If it isn't, it *should* be," my mother said. "Sounds like the island where this one grew up." She gestured at me by tilting her head as she drew on her cigarette. We had to wait for her to exhale before elaborating. "Big tourist trap nowadays, but back then was a haven for fugitives, too. Our dentist was fleeing cocaine charges in Boca Raton—"

"What? There's a place called that?"

"Yeah, just south of here."

"Oh, God. You know what that means, right?"

"Yes, dear. And we call it that amongst ourselves. Same with my nephew's wife from Louisiana, we refer to her home as Red Stick."

"Ahh," Elizabeth said, smiling and tilting her head back, the smoke curling away into the blackness next to her face. She had crooked teeth that I tried not to stare at. One of the front lower ones was set back far enough from the rest to be a hazard later on, when we were alone, and that was the first thing that I'd noticed that went on the Cons list, across from a *lot* of Pros. "So, wait a minute, let's get back on track. You're taking us to a bar?" she asked me.

"Yeah, there's a few on Clematis, downtown, but I only really like one of them. It's a wallflower bar, a total dive, and all the people that can't dance anywhere else can do their awful moves with impunity there, without fear of reprisal."

"Fear," she said, grinning. "Of *reprisal?*"

"You making fun of me?"

"I haven't decided yet."

"You'll find I use a lot a them ten-dollar words."

"Ahh," she said, nodding. "Is *that* what you call them?"

"Sure. Why use ten one-dollar words when you can use just one ten-dollar one?"

"Hadn't thought of it that way."

"Funny, I hear that a lot."

"Do you, now?"

"I do indeed."

Charlie cleared his throat. "Get a room you two."

Elizabeth grinned her charmingly crooked teeth at me, and my mother blew smoke at us, asking "You'll all be coming back to the apartment, then?"

The apartment she and I shared. Not because I wasn't fit to live on my own like a grown man, like she said, but because if I wasn't there to keep an eye on her, she'd drink and smoke herself to death. Or maybe we were both right. Ehhh, either way.

"Eventually," I said, blowing a smoke ring and watching it unravel, then meeting Elizabeth's eyes. She smiled somehow without actually smiling, cutting her eyes quickly at her boyfriend and back to mine and narrowing them slightly, and our campaign to get Charlie shit-faced began that very moment.

Now, Josh was a bouncer at Respectables on Clematis. And for a bouncer, he wasn't that big and was just as short as me, but he showed me something once about being smarter than the other guy that I respect and admire, even though I can't do it myself and probably wouldn't even if I could.

Josh had *really* long blond hair, but I always forgot about it, because he wore it in a pony tail and never took off his black baseball cap, and the sides of his head were shaved bald. Had one of those goatees minus the moustache too, and a big ol' nose ring like the bull had that time Bugs Bunny went to Mexico. Or was it Daffy?

It was that nose ring that kept the two of us from getting in a brawl up in Dr. Feelgood's one night, with a big surly jackass who was running his mouth to his big jackass friends. See, my problem is I'd been small and bullied all my life, until I started going to the gym and drinking shakes and popping pills, and I have maybe got a teeny tiny chip on my shoulder about it. And I look pretty good, and have a bit of a swagger that people mistake for arrogance that's really because of a hip injury years ago...or maybe it's part arrogance, part injury...or maybe I made that injury up years ago as an excuse to walk with a swagger. I don't really remember, it's been so long. Either way. So some people, usually in bars, decide I'm the guy that's to blame for whatever their problems are that night.

So I'm standing there, minding my own business, with Josh and his nose ring, while Cochise over to my right is talking about how pissed off he is to the other six people in his group, three huge men and their wenches. And blah blah blah I'm fixin' to go *off*, man. I just wanna fuck somebody *up*. Like these two faggots over here.

And all seven faces turn towards me and Josh.

Now, Feelgood's is a club, and music was blaring, but I learned how to read lips a long time ago, and no matter how many times my Mom says I'm just paranoid and I can't read lips and I just made that up about what that guy over there just said, I can read lips just fine, thank you, and yes, he called me a faggot while I was reading his lips reflected in the bisected mirror-backed shelves behind the bar.

That's not the problem. People say I am gay all the time

because I'm better looking than them and dress well. I've had time to get over it. The problem was I'm just minding my own business and this wannabe alpha male wants to put down on me and ruin my night.

"Did you hear that?" I asked Josh, with my voice sounding maybe a little too eager.

"Yyyyep," he said wearily. So. I wasn't mistaken.

Since I wasn't watching him, I couldn't make out what the guy was saying, but he was getting a little louder, as if in the hopes that we might hear. Josh turned his head ever so slightly in his direction and cut his half-closed eyes at him while pushing away from the bar. I must've changed my posture somehow because he said No, hold on. Watch this.

He turned to give them all a good view, but not a challenging one, while he handed me his drink and started unscrewing his nose ring.

"He's taking out his nose ring," said the jackass to his buddies. "Oh shit. He's taking out his nose ring. He's taking his nose ring out." Like they *needed* the play-by-play. "I'm going to stand over here."

All of a sudden his rage at whatever had drained out of him and he was chickening out. And in front of three women, no less. He went around his buddies to put them between us. I couldn't believe it.

It took a minute for Josh to get the ball undone, slide the ring out of the huge hole in his septum, and pocket it. Then, "Don't move. I'll be right back."

I watched him walk over to those Neanderthals and their wenches, push his way through them, put his chest up against the jackass's belly, and hold the back of his hand up in between their faces.

"Hi, I'm Josh. I work here. You having a good time?"

The huge guy's wide eyes flickered from fear to relief, and he grabbed Josh's hand and squeezed it.

"Yeah! Yeah, we're having a great time!"

Meanwhile, one of his friends, the biggest one, was glaring

at me like I had run over his dog. The rest of them were looking at Josh, but Gigantor was trying his damnedest to stare me down. Like kids on a playground. I just stared back at him, unblinking, with a hint of a smile, while Josh and Cochise or whatever his name was became the best of friends. Josh let go of the asshat's hand and slapped him hard on the shoulder, told him that it was great to meet him, then came back over to where I stood stupidly holding two drinks.

The other guys were quiet for a second, until the loudmouth said awkwardly "Let's go over that way."

And away they went, the big bad mothafuckas who take no shit from nobody, and their wenches. And I could not stop laughing.

While putting his nose ring back in, Josh quoted me a line from *The Godfather*. The book, not the movie.

"There are men in this world who go about demanding to be killed. They argue in gambling games; they jump out of their car in a rage if someone so much as scratches their fender. They wander the streets calling out 'Kill me, kill me!'"

"Yeah, I know, and that guy's one of 'em," I said, sneering and watching them watch me from afar.

"No. You are."

"*What?*"

But he was drinking his Jack and Coke, and he couldn't elaborate.

Now, reason I think why this worked, same as this time I was in college and I had a mohawk. I'd just gotten two flat tires driving down a side street in the middle of the night, and had walked to the corner of the nearest more beaten path to wait for a tow truck driver. I was wearing a baseball cap backward, sitting on the sidewalk and leaning against a lamppost, when this car full of assholes pulled up at the stop light next to me. I guess I had a sign on that said Fuck with me.

"Hey, man!" One guy calls. "You got a cigarette?"

"Nah," I said. "I was hoping you'd have one."

Now, keep in mind, I was a hundred ten pounds, soaking wet. Skinny as a rail.

He muttered something I couldn't hear, and one of the girls in the car said "Careful, he'll hear you!"

"Pshhh! I don't give a shit if he does, I'll kick his ass!"

So, like anyone would, I bolted to my feet, whipping off my hat, and ran at the car. It was exhilarating, the way they all screamed and squealed their tires running the red light and disappearing down the road. Grossly outnumbered, I chased them off with only a bad haircut.

It's like those nature shows where you see a small fish or an insect with patterns on them that look like great big eyes. A huge predator comes along and the little guy flashes these "eye spots" at the big one, making them go, in this quick, unthinking, panic inducing moment, "Oh shit! This thing is *way* bigger than I thought!" And they book. It's just like that, I think. People aren't so quick to fight some little dude with a bigass ring in his nose or a bad haircut because he must have something up his sleeve, because it takes a different kind of guts to weather the disapproving stares of everyone else you meet, day in, day out, everywhere you go.

Yeah, so anyway, Josh was standing outside of Respect's with his hands in his pockets and leaning his back against a palm tree growing out of the sidewalk. He was wearing his Heath Ledger shirt again (he either had ten of them and he wore a different one every night, or he washed it every morning—either way, it was always clean) and talking to one of the socially inept that had to stand outside and talk to the bouncer. At a wallflower bar, no less.

The guy talking to him was a tall fat website designer in a Decepticons tee shirt, who gestured towards us and pushed his glasses back up. I could never remember the guy's name, but wasn't bothered about it either. Josh looked in our direction, and sized up my companions before smiling and tilting his head backwards at me. Somehow, his arms looked bigger than they had the week before.

I made introductions, yadda yadda yadda, we went in and got a booth, and I went to the bar feeling a little self-conscious in my suit, but played it off by pompously slapping the bar with my palm, shouting "Bottle of your best champagne!" at the bartendress, who had more tattoos than I'd ever seen on one person. She laughed and held up a longneck, putting on a French accent and sneering "Le Boodweisair."

I got us four Cuba Libres instead, one extra in case Josh wanted to sneak one while on duty, and Charlie immediately started annoying the shit out of us.

"Yunno, I'm a bit of a DJ myself. You think that they'd let me spin for a bit?"

"Um, no. I don't. They have a guy here they pay to do that."

"Yeah, I know, but why don't you ask him? Ask him if he'll let me."

"I can already tell you what the answer'll be."

"Yeah, but—"

"Dude."

"I'm really good at it."

"I'm sure you are."

"Don't patronize me, man. I just want to spin for a few minutes. I'm really good. Why don't you ask him? Go ask him. Please?"

I felt something touch my leg and looked over at Elizabeth. She was slumped down low in her seat and eyeing me the way evil people do in movies. I tried not to smile at her in front of Charlie, and squirmed a little in my seat so she could reach me better with the foot she'd slipped out of her flip flop.

"Sorry, Charlie. I really don't feel like leaving my seat right now. If you want to go ask, be my guest."

"But I don't know him."

"Hey, I don't know him either. If it means that much to you, go talk to him yourself. But drink up, before your ice cubes melt."

He looked down at his drink as if he'd forgotten he'd had it, and his eyebrows jumped. Elizabeth's foot was groping blindly, and I tried to steer it into place with my thigh while her jackass boyfriend drank until the glass was drained, swallowing loudly, smacking his lips and giving a loud and obnoxious sigh of satisfaction.

"So," Charlie said. "You never asked what I do for a living."

I blinked at him for a second, wondering if I had heard him right, then glanced over at Elizabeth, who was rolling her eyes again without actually rolling her eyes. So, yeah, apparently he really had said that. I blinked at him a few more times, shook my head a little as if to clear it—a little move I'd picked up from my father that really pisses people off—cleared my throat, and asked him "What?"

"You never asked," he repeated. "What I do for a living."

I nodded slowly, chewing on the little knob on the inside of my cheek, and furrowed my brow in concentration.

"Hmmm. You're right. I haven't."

Elizabeth's toes were as close as they could get to where I wanted them to be without me readjusting myself in my seat, and I just couldn't do it with this idiot's eyes on me. Frust. Rayyy. Tingly close.

"So?" he asked.

"So, what?"

"Aren't you going to ask me?"

I sighed. "What do you do for a living?"

"Guess."

And I swear to God, I almost came flying across the table at him but Elizabeth's cute little toes lunged forward those last two inches and I tensed up all over. It was so blatantly obvious looking at her now, she was slumped that low in the couch. Hell with it, I thought. I really didn't care anymore. Actually, I kind of wanted him to see it, but it was Josh who ended up noticing when he brought us all shots from the bar, his treat.

He was smiling and just about to say something when he glanced down, and his eyebrows jumped at the small bare foot digging into my lap, and he started clearing his throat and shaking his head, trying not to laugh. Charlie was looking at the shots as if they were an interruption instead of a gift.

"What're these?" I asked.

Josh coughed and hit himself in the chest a few times with his fist, cleared his throat one last time, and gave me a questioning look. I gestured with a jerk of my chin at the shot glasses and he said "Try one."

"Yeah but, what are they?"

"Just try it."

"What's in it?"

"Try it!"

"Okay."

I distributed them among my guests, crossed myself Catholically, and downed mine. It was a bright red something-or-other that tasted really familiar, but I just couldn't place it. I asked again what it was, and he said Pineapple Upside-down Cake. I frowned, then nodded, and shrugged. Yeah, okay, I could see why.

"So, Josh," I said. "You'll never guess what ol' Charlie here does for a living."

Josh sat down next to me and looked across the table at Charlie, squinting.

"Chimney sweep."

"No!" Charlie said, slumping. "Do I *look* like a chimney sweep? I don't even know what a chimney sweep *is!*"

"I know," Josh said, snapping his fingers. "I've got it. You wash people's windshields with a dirty rag on the street, when they don't want you to."

"No! No, I *don't!* Elizabeth, tell them what I do."

"I don't even know."

"Yes, you do!"

"Fine. He's lying. He *is* a chimney sweep."

"What the fuck!" He looked around as if searching for the

hidden camera.

I asked Josh if the DJ might let Charlie here spin a little.

"You kidding? Of course! Let's go, champ." Josh got up and waved his hand for Charlie to come with him. Now the guy looked like he wasn't so sure.

"Oh, I dunno. I don't want to do it now."

"Aw, c'mon! Don't be such a wuss."

"I'm *not* being a wuss! I'm just not in the mood anymore."

Elizabeth took her foot out of my lap and sat up. I tried not to show my irritation.

"Go, Charles. Show us what you can do."

"No, I don't want to."

"*Char*les."

He wavered a little, then gave up. "All right."

Getting up, defeated now, he followed Josh half-heartedly to the DJ's booth across the empty dance floor. Elizabeth watched him go and leaned across the table to hiss conspiratorially at me, "He is *not* my boyfriend."

I smiled indulgently at her, and she insisted.

"He isn't! My grandmother just thinks he is."

Putting my hands up as if I didn't disagree with her, I said "Okay, okay. If you say so."

"Good. So I'm trying to decide how to pursue my attraction to you. I'm leaving, going back to Antigua tomorrow, but I'll be back in May."

I nodded, trying to pretend I wasn't wondering what the actual question was.

"What do you think?" she asked.

"I think..."

Stalling, I glanced across the club at Charlie and Josh arguing with Patrick, the DJ, and watched them for a moment. Josh was making a big show out of it, and Patrick could tell it was a big joke, and was playing along. Charlie was pathetically earnest.

"I think we need to get that boy passed out."

"Got any weed?" she asked. I looked at her in surprise.

She didn't seem the type. Not like there really is a type, per se, but she seemed more the type that Didn't.

"I haven't had anything to smoke in a whole week. I have been cooped up in my grandmother's house, and I'm going to just *die* if I don't get high tonight."

"Um, I don't smoke anymore, but Josh can hook us up."

"Good. You'll talk to him, won't you? I've a pretty high tolerance, but Charlie will be out like a light in no time, especially if we can get him drunk first."

"You mean he's not drunk already?" I asked, incredulous.

"Oh no," she said with a sneer. "He's always like this."

"You haven't slept with him, have you?" I asked, suddenly seeing her social value plummet if she had.

A look of resigned embarrassment said it all, but she mustered a little dignity.

"I don't know. I got drunk once and woke up in his bed. He said we did, but I had my panties on, so I don't think so."

I tried to decide whether it made a difference, and then remembered that it didn't. There's a grace period when you've been drinking, even on a weekday, where your social value is unaffected by sleeping with a girl who's slept with a dork. I think. If not, there should be.

Josh came back over to us without Charlie and I asked him where Elizabeth's boyfriend was.

"He's not my boyfriend!" she said. Josh smirked.

"He went to the bathroom. Want another shot?"

"Yes please," I said, trying not to laugh, and he looked at Elizabeth.

"I'd love one," she said.

"Okay. I'll get your boyfriend one, too."

"He's not my boyfriend!" she screamed, banging her fist on the table.

I watched that recessed lower tooth while her mouth was open, thinking "Man, that's probably going to hurt me, later."

And yeah, it did, to an extent. But it was somehow tolerable, dulled just enough by several hours of drinking and the little bit of smoke I actually inhaled when Josh came through for us. The rest of the times I took a hit off of the joint Elizabeth'd rolled, I was blowing on it to make the cherry light up, and pretending to hold the smoke in so long that it stayed in my lungs.

Good ol' Charlie though, oblivious, smoked away and kept remarking that, damn, that was some good weed, which it wasn't. It was stuff you get on short notice from a college kid on a Wednesday night, but it was doing the trick. He was fading fast.

What worried me though, was that Elizabeth was smoking just as much as him, if not more, and it seemed she might pass out too, making the whole thing a waste of time and effort.

Then, at one point, Charlie asked if I was gay.

I blinked at him again, glanced over at Elizabeth, who also didn't know what to make of it, and asked him what made him ask me that.

"I dunno. I just thought…maybe…"

"Because I am."

They both looked at me in surprise.

"What, is that going to be a problem?" I asked.

"No!" Charlie said. "No no no no no!"

Elizabeth had a sly little grin spreading slowly across her alabaster face.

"I try not to make a big deal out of it, though. I don't go around making sure everybody knows it."

"That's good," Charlie said, then corrected himself. "No! It's not *good*, but it's better than, I dunno…"

"No, it's okay," I told him. "People really should not be going around telling everybody what they do in bed. You don't hear people saying they've got a foot fetish or they only get off when they stuff their noses full of raisins, as soon as they're introduced. Nobody ever tells me they're heterosexual when I ask them to tell me about themselves."

"Yeah, that's always bugged me, too," Elizabeth chimed

in. "People are quick to tell me they're gay right off the bat. Do I volunteer my bisexuality? No. Seen it on The Real World and those shows, too. Day One, some guy moves into a room and introduces himself to the girl sleeping next to him and says 'By the way, I'm gay, I dunno if that's going to be a problem for you.' Like if it was, she was, one, going to say it on national TV, and two, going to say it on the first day as soon as she meets her room-mate, just volunteering herself to be the first one that's voted out of the house."

"And why would it matter unless he *practiced* his homosexuality with someone else in the bed right next to her?"

"Or any kind of sexuality, for that matter?"

"Good point. That kind of thing should be done in private, no matter how you choose to do it. So, now I have to ask, why are you asking, Charles?"

His posture had completely changed into that of a timid, submissive, and very vulnerable young man debating whether he should come out of the closet. As if this were the opportunity to admit it to himself and maybe explore this exciting new facet of his personality with a gorgeous young man. He looked at the floor.

"Oh, I dunno..."

Good. Now he wouldn't worry about me being left alone with his alleged girlfriend, and I had him on the defensive. He might even have to go off, sit alone somewhere and think about a few things for a while, giving me time to at least make out with her again.

Oh yeah, forgot to mention that. I had gone into the Men's room earlier and she'd come in after me. No preamble, just straight to the kiss, until good sense prevailed and I told her we should go into the Ladies', where Charlie wouldn't stumble upon us. She nodded, and I stuck my head out the door to make sure he couldn't see us, then ushered her into the other bathroom, pushed her into a toilet stall, and took her breath away.

She even fanned herself when she pulled away from me,

which is a reaction I've always enjoyed. It was nice to finally see a little color in those cheeks.

So, we finished smoking the joint out by Josh's truck, parked in the alley behind Respect's, and pushed our way back through the fans of palm fronds and ferns that had obscured us from the road. My eyes lit up when I saw Curtis ambling down the sidewalk toward the bar, and for the first time in my life, I was happy to see him.

Thing about Curtis, see, he was always telling me about the girl he did this and that with last night or earlier that day and making gay jokes about everyone else in the room, every now and then even insinuating that I might be gay as well. Perfect.

He just had to meet Charles. He just *had* to, like, omi*god*.

Curtis and those mango mojitos he insisted on drinking. And he'd drink so many of the goddamn things that he would start to talk like that. Over-enunciating and saying "Omigod" and then trying to bite back his words. Trying to cover it up with "Oh Jesus, look at them faggots over there."

I only saw him when I was drinking at O'Shea's, across the street from Respect's, trying to put off going to the wallflower bar as long as possible, with the others that were reluctant to go but would end up going anyway. Then Curtis would show up and be Curtis at us until we all drifted out and sought refuge with the Goths and Punks.

"Curtis!" I stage-whispered. He looked at us in surprise, like "Omigod, somebody is talking to *me*." I gestured with a slight upward tilt of my head and he came up to us, wearing his Castro hat that only looks good on chicks. Or Castro.

"Wassap, Curt?" I asked, doing the Handshake with him, the one with the thumb squeeze and finger-snapping that took me forever to learn (but don't tell anybody).

"Lookit you all dressed up! Where you been at?" he said. Then he interrupted himself. "Duuude, I can barely *walk*, man. I am worn the fuck *out* by this bitch I just rolled off of. Crazy, too, man. We did it like dis—" He started doing a little dance.

"We did it like dat—"

"Did you do it with a whiffle ball bat, Curtis?"

He stopped and smiled, clapping me on the shoulder.

"Yeah, dog. You caught me. I was just listening to that on the way over."

"Classy, man. Classy."

"Ain't you going to introduce us to your friend?" Elizabeth asked, a very subtle tone in her voice hinting that she was severely judging me for knowing this guy and even being on a first-name basis with him. It did not seem to matter to her that Curtis didn't know *my* name, though.

I introduced them, and noticed the way Charles and Curtis looked away when they met each others' eyes. Both of them looking down. And looking back up. Perfect.

"So, what're you doing?" Curtis asked me, looking back over at Charles for a split second.

"Well, we're going to Respect's," I told him. "Just burned one out back."

Elizabeth and Charlie both looked at me sharply in alarm, but I said "S'okay, he's cool," which bumped Curtis's self-esteem up a bit and made him feel like he could be a part of something. Of *Us*, maybe.

Which, by Elizabeth's ever so slight hunching of her shoulders, she did not want happening under any circumstances. I wished I could give her hand a reassuring squeeze, tell her I knew what I was doing, but couldn't. It would ruin everything.

"Yeah, yeah, listen, doncha worry about nothing. Like the man said, I'm cool. Hey, so, you got any left?"

"Yeah, man, but later." I told him. "We're good and tight right now. So, you wanna hang with us?"

His eyes lit up like a puppy dog's.

It was pathetic.

"Yeah, man, sure!"

I told him it was a pretty big night for us, that we were celebrating something, while we walked the rest of the way to the door where Josh and Whatshisname stood, smirking. I forget how I phrased it when he asked me "What?" I know I said something back that sounded like Salinger, like "a momentous reversal of mental polarity" or "momentous change of polarity in our minds" or some shit. Something with "momentous" and "polarity" in it that made me sound intellectual and had Elizabeth raising her eyebrows at me. "And we need to drink a little more so we can let it settle."

"Let what settle?"

I held the door open for Elizabeth and Charles, the music tumbling out onto the sidewalk. They went in, and I had to shout a little for Curtis, Josh, and Whatshisname to hear me.

"We all just came out of the closet."

The look of stunned disbelief on Curtis's face was priceless. I glanced over at the other two. The Decepticons guy was staring, dumbfounded, but Josh was choking laughter into his fist and turning away.

"So, you coming in?" I asked, stepping through the door and letting it swing slowly behind me. Since Curtis was standing close to me when I said that, leaning in so he could hear me, the door came to rest on his shoulder, and was held open by his weight. While I walked into the darkness I watched the light cast on the floor by the streetlights outside and knew he was still standing there, adjusting his mind to this revelation, deciding what to do.

Does he back out and go to O'Shea's, continue his self-denial and be miserable, or take the mental step everyone else was waiting for him to take so he could stop being a freaking laughingstock? He certainly couldn't stand there in the doorway for the rest of his life.

I could picture what it must've been like for him. With Josh and the Decepticons nerd watching him, he couldn't just walk in. It would've been like admitting to them that he wasn't disgusted by what he had heard. And then they might think that *he* was queer too. Omi*god*. But he could stay out there and bluff it out, playing his role for the benefit of two people he didn't even know, maybe more for himself than them, really. And stay in the lights and the metaphorical sanity they might represent, and the loneliness, or plunge forward into the darkness and the life of new possibilities and company of others.

The door opened wider, and a long purple shadow came through it. I turned and looked, but it was Josh. Josh with his nose ring and a gigantic grin splitting his face underneath it. Shit. Well, I'd taken a gamble. And, like all gambles when you come right down to it, the odds weren't good.

Josh covered his face with his hands, and his shoulders quaked while he walked toward me. The strobe and the colored lights that flashed across him made him move in jerky motions through the darkness. He got to me and flashed his grin again, in red and yellow and purple light.

"I know what you're doing, Johnny-boy," he said. "You shoulda waited til he was inside. Then he wouldn't be able to walk away so easily."

"Doesn't matter, there's still my Plan A working for getting rid of Charles. Chuckie."

"Is the door still open behind me?"

I looked without turning my head. "Yep."

"Then, hey," Josh said, pushing past me. "The door is still open."

We left Curtis and Whatshisname out there to simmer in their little stalemate of personal cowardice, both wanting to come into the club but neither wanting to admit in front of another that he was willing to go into a place where gay people were. Neither would get accepted anywhere else, especially not O'Shea's, but most people would rather be dorks than queers. Well, maybe not for much longer, what with the zeitgeist and all, but for now—

Hold the presses.

The door came open again and swung shut behind whoever had entered. I was following Josh to the bar where Elizabeth and Charlie were, and didn't want to look over my shoulder and scare whichever of them it was away, make them lose their nerve, and had to square my shoulders and made a concentrated effort to keep myself moving forward. The temptation was that great.

I had a flash of inspiration, and decided to stop and light a cigarette so whoever it was could get to me before I reached the others, and speak to me alone if he needed to. Didn't matter how loud it was in there, the proximity of others would make him chicken out. Another thing I'd learned, the proximity rule: you can't cramp someone's style with other people.

I felt a light touch on my shoulder while I was puffing my cigarette to life, and turned to see Whatshisname pushing his glasses up. Shit. Oh well.

"Listen, you were just joking out there, right?"

I tried hard not to smile, but the corners of my mouth twitched. "Nope. For real, we all had a major breakthrough. Charlie over there, he came out, and then me and her, we admitted it out loud, and it was really beautiful. It's like going to confession, you know? A tremendous weight's been lifted off of us and I feel, for the first time in my life, really, actually, *free*. You know? Like, I don't have to lie to myself anymore."

I watched him nod, his face still with this mask of studied revulsion, and I could see wheels turning inside his head while he digested this.

"And you know what?" I said, deciding to ice the cake. "I feel like dancing. I really don't give a shit if I have the right moves, either. I don't give a shit what anyone thinks about me right now. I'm gay and proud of it. I'm proud that I have the courage to say it."

And I left him with that, that little seed planted in his head, and went to the bar trying my damnedest not to laugh. Yeah, I'm gay, I thought. I'm a great big dyke.

"What're you grinning about?" Elizabeth asked.

I looked at Josh, who covered his face again.

"Aw, nothing."

"So, anyway, back to what I was saying before I got so *rudely* interrupted. We need a book for all the words that aren't in our language. Like, you know how Germans have all these concepts that they condense into one word, like *shadenfreude*? Well, all we do in English is just borrow those words instead of translating them. We call it just *shadenfreude* instead of—"

"What's *shadenfreude*?"

"It's when people are delighted by seeing other people's misfortune."

"Oh, like the audience in Jerry Springer."

"Yeah, and all the people who'll watch videos of horrible accidents, like people who just *have* to watch things die, but from a comfortable distance. From the safety of their couch in their living room, they have to see people on safari torn apart by wild beasts on the Serengeti plain, or gazelles getting pulled down by crocodiles into watering holes or water-skiers being crushed by speedboats on camcorder video clips that people have no problem getting paid to air on national television, regardless of the grief of the dead peoples' families."

"Or that asshole selling copies of the Columbine security footage," I said, nodding. "But, so, when you come up with the word for this, in English, that's not a loan-word from some other language, how are you going to construct it? With Greek or Latin roots, like just about every other word we have?"

She paused, seeing where I was going with this train of thought, and wrinkled her nose at me.

"I get your point, but ya don't have to be a dick."

"I didn't mean to be—" I was about to say, but then I remembered that rule about apologizing. You don't do it. So, I shrugged instead and took a sip of her drink. Another thing I learned, bad manners can be overlooked, but cowardice, never. And the definition of cowardice is so broad it encompasses everything right down to "being quick to make sure that you haven't offended someone."

The idea is, if you want somebody to like you so much

that you won't risk their not liking you after misunderstanding something you've said, then you must not be secure in your own belief that you are the rightful king of wherever you are. And if you aren't king or at least warrior-chief potential, chicks are evolutionarily hard-wired after millions of years of evolution to kick your ass to the curb and find someone else who is.

Therefore, I chose to appear a dick instead of a pussy, and she continued talking.

"So since you seem to have a pretty good turn of phrase and a strong vocab, I want you to help me compose this dictionary."

I looked at her in surprise. Couldn't help myself.

"First thing's first," she went on. "I want a word that'll encapsulate what Jeffery Eugenides meant when referring to 'the hatred of mirrors that comes with middle age.' And go ahead and use Greek and Latin roots if you have to."

"But it's going to turn into another one of those bullshit disorder titles," I protested. "Like Attention Deficit Disorder, or Borderline Personality Disorder. It'll be Self-Image Disappointment Due To Advanced Maturity Disorder—"

"Yeah. Now make it one word."

But I ignored her, pretending to go off on a tangent rather than admit that I had no idea what to call it. "And that's something that pisses me off. We're told to be comforted in that we're unique, but then the very differences in us are catalogued and labeled with euphemistic pomposity, followed by the surname of Disorder. Intermittent Explosive Personality Disorder. Post-Traumatic Stress Disorder. As if whatever the "proper" Order is, that's what we should strive to be through medication and counseling—to what end? To all be *alike?* Uniform? It's okay everybody! You don't have to be unique anymore! There's a McDonald's in every country now, and a one-night-stand in Rio de Janeiro is no different from masturbating at home! Hooray! Only one flavor jellybean from now on!"

And, thank you God, before she could point out what I was saying made no sense whatsoever, Whatshisname came up behind me and put his hand on my shoulder, leaning in between us and interrupting.

"Y'know what? Fuck it. I'm gay, too."

Elizabeth's mouth dropped.

And, like a CD skipping, she completely forgot about being frustrated with me.

"Great!" I said, turning on a thousand-watt smile. "But yunno who really ought to hear about it?" I jerked my thumb over my shoulder at Charlie, who was arguing with DJ Patrick again. "Him. The ring-leader. Our kind of, like, father."

"Yeah?"

"Yeah. He'd love to hear it. There's safety in numbers, yunno. It'll give him courage. *We give each other courage.*" And I gave him a look with all the false significance I could muster.

When he'd gone off to announce himself to our Fearless Leader, Elizabeth asked me "What in the blue fuck was that about?"

"I'm getting rid of Charlie. Soon he's going to be off singing karaoke in a gay bar somewheres and we're going to have the rest of the night all to ourselves."

She stared at me while I drank the rest of her drink and called for two more.

"Johnny?" she said, and I raised my eyebrows at her. "I don't ever wanna get on your bad side."

I smirked, and glanced over at the DJ's booth where Charlie was looking at Whatshisname like he had antlers growing just above the glasses he kept pushing up. The hapless web designer seemed to flounder a little, and both of them looked over at me simultaneously in unison, at the same time. I gave them an encouraging little wave and a smile, and turned back to Elizabeth. I was about to say something when I saw Curtis standing in the corner of my eye, illuminated in staccato flashes by the lights slashing across him.

He was looking at me as if working up the nerve to say something, and then Charlie was at my side.

"Johnny, uh...hey, listen."

"What's up, Charles? I'm all ears."

"Well, this guy over here, he's saying some weird shit."

"It happens, Charlie. We're in a bar."

"No, I mean, he's telling me he's gay."

"It happens, Charlie. We're in a gay bar."

He gaped at me. "We *are?*"

"Um, yeah. I thought you knew that."

"What? You never said anything!"

"I didn't think I had to. I thought that's why you were asking me earlier, to make sure. What's the matter? You want to leave?"

He opened his mouth and closed it at me a few times, considering it. "No, I guess not, but it is a little weird, yunno."

"Look, you're perfectly safe. What happens in here stays in here, and it happens here all the time."

Actually, no. It doesn't. Funny story (at least to me) a week before, on a Thursday, I was making out with this chick Christa, and everybody was hootin' and hollerin' for us to get a room, and then the next night when I was making out in the same spot with two chicks in this little group effort, nobody said a word. They just watched. And some of them had their phones out, filming it. Sickos. But the point here is, if it happened all the time, what was all the fuss about? And why were people talking about it in O'Shea's on Saturday?

"*Does* it?"

"Indeed, it does. Why, right over there on those couches, you'll see people of the same sex making out. Last Friday, a couple of girls were going to town on that one in the corner." I glanced over at Curtis, who was looking at us uncertainly. I picked up one of the drinks that had arrived when I wasn't paying attention, and held it out to him. It was like convincing a squirrel in the park to come closer and accept the peanut you held out for him. But after a moment's hesitation, he came forward and took it, had a tentative sip, and then downed the whole thing in one go.

When his eyes opened again, he was looking at Charlie,

and Charlie was looking at him. I picked up the other drink off of the bar and nudged Elizabeth with my elbow, jerking my chin at the table we'd been sitting at earlier, and she followed me there.

"Here's another word we need to come up with," she said. "The revulsion you get when you can read other people's minds, and it's unattractive."

"Yeah, I know. I'll get right on it. In the meantime, your grandmother was saying something about my painting your portrait. We're going to have to take a few reference photos before you leave."

"Oh God, no. Right now? After I've been drinking?"

"You look fine to me."

"That's the rose-colored glasses from all the rum going to your head."

"Or your perfume. It's been messing with my head ever since the bathroom."

"I'm not wearing any."

"You're kidding. That's how you always smell?"

"Well, it's not *perfume*. It's a little something I mixed up myself to enhance my pheromones. I make a fragrance out of macadamia oil, essence of jasmine, and non-menstrual vaginal fluid, and dab a little behind my ears."

I blinked at her a moment, then glanced back at the bar to see Curtis and Charlie talking. For the first time that night, I was completely at a loss for words.

Sensing this, she just picked a topic and started running with it. I didn't pay attention at first; my head was buzzing with the sudden realization of all the weirdness of the evening that I had just accepted, without actually considering. Elizabeth was talking, and I nodded and made appropriate facial expressions whenever I instinctively knew I should—a skill I'd picked up a long time ago, and I'm not really sure how. Eventually, I decided that this chick was worth it, all this trouble, because she was a helluva lot more intelligent than Tiffany, my last girlfriend, or any of the girls I had cheated on her with, and I ought to keep this campaign going.

I tuned back into what Elizabeth was saying, and was startled by it.

"—uncertainty about sexuality, followed by the realization that you're different from everybody else, I think, forces disidentification from socially conditioned thought or behavior patterns, that automatically raises your level of consciousness above that of the unconscious majority. I think being an outsider makes life hard, but also places you at a bit of an advantage where enlightenment is concerned."

"Maybe," I said. "But it also causes a developed sense of identity based on that sexuality, or whatever, and you end up playing roles dictated by this mental image you have of yourself as whoever you decide to be that week."

"Wow. I didn't think you were paying attention."

"I wasn't. I just winged it. Wung it. Whatever the past tense is. Was I on the mark?"

"Close enough."

"Cool. I'll forge boldly ahead, then. I say you get your sense of identity from things that ultimately have nothing to do with who you are. Your possessions, your appearance, your achievements and your failures, your social role, belief systems, yadda yadda yadda. And this false, artificial, mind-made self is just hollow and insecure that it's always looking for new things to identify itself with. Then we've got these rituals to remind ourselves of who we are, as if we need to be told. Then only fear and neediness are left after the feeling ends."

A bunch of bullshit, probably, but she seemed to be eating it up. I didn't want to continue in that vein, though, because I didn't like the way it made me look at myself. Like maybe I wasn't an artist or Casanova or a playboy. Like maybe those were all things I had decided to be so that others would see me standing apart from the rank and file.

Then I remembered why I'd stopped smoking weed all those years ago. All of these stupidly "profound" introspective journeys that I had subjected myself to, that invariably ended in mental masturbation conversations like this. Maybe I was higher than I thought. Damn. I made a mental note to never even pretend to smoke weed again.

"Yunno what I'm thinking?" Elizabeth asked me, smiling without actually smiling.

"Yeah. You're thinking Damn, not only does this guy understand the things I'm saying when I subtly test his intellect, but he also has something lucid and informed to say back." Her eyes went as wide as dinner plates, and her smile bared every tooth in her mouth.

"You know what else we're going to need a word for? Those shivers you get when someone vocalizes your internal dialogue!"

I smiled, pretending to blush, looking away shyly, while secretly thinking "What?"

Now, yunno when you're drinking maybe a little more than you should, and you time-travel? I know I'm not the only one this happens to. You have even so much as one more sip of a drink, and next thing you know you come to and you find yourself walking with a bunch of strangers down a dark street you've never seen before. Or you're on a yacht somewhere with a couple of people a little older and richer and drunker than you, and some woman with poofy hair says we ought to take off all our clothes and go swimming. Or, on those lucky nights, you're in your car, finding out you're right, that one recessed lower tooth does hurt, and it scrapes you in a way that you know is going to be on your mind for the next couple days until it heals, but you don't have the heart to stop her.

Well, the last thing I remember before taking one more sip of Elizabeth's drink was doing a double-take at Charlie and Curtis, when I happened to glance over, and they were...um ...well, they weren't just talking anymore. So, yeah, you can imagine I'd take a drink right then. And *boom!* the flux capacitor fired up and I fast-forwarded to the backseat of my car.

Which, in a way, kind of pisses me off because I'd like to

know exactly how I got there, what way I touched her, what I said or how I said it, whatever it was that got us to get to this point right here. Yunno, so I could *remember it?* And maybe, I dunno, do it again sometime?

Sometimes I think it would be nice to have a guy with a camcorder following me around when I drink, but then I remember that one time in college when I did. And I was a little horrified when I saw the tape the next day. Apparently, I have a temper.

Anyway, enough was enough. That tooth was killing me.

I grabbed Elizabeth by her Aeon Flux hair and pulled her head up and back, leaned in and kissed her roughly. She adjusted herself and I helped her lift her dress up over her head. It was amazing how skinny she was, and pale. It made me feel guilty, in a way, like I was taking advantage of a child when I put my arms around her and tried to hold her tiny, frail, bird-like body.

I couldn't help thinking of the withered skeleton of a fallen leaf, or the tattered wing of a dead beetle. Something so light and thin that it would crumble to dust if I fucked it too hard. Her breasts were too small for her to have need of a bra, so she hadn't bothered to wear one. All there was to do was curl a finger under her panties and yank them aside.

"Waitwaitwaitwait. Condom," she said.

"Jacket pocket," I said. "The small one on the left side way down low."

She reached over into the front seat and fished her hand around in my jacket, eventually pulling a long strip of them out and tearing a crinkling, shiny square off of it.

"You think you have enough?" she asked.

"Only one way to find out."

She tore it open and was fiddling with its contents when a sudden rap at the window startled the shit out of us.

"Sorry," Josh whispered through the steamed-up glass. "Ya gotta do something bout yer boyfriend."

"He's not my boyfriend!" Elizabeth shouted.

"Whatever. He's Curtis's boyfriend now, too, and we can't have that going on in there. This is a family place."

"Family place, my ass," I said. "You weren't saying that last week."

"That was different, man. That was two girls."

"Double standard!"

"They call it a double standard cause it's twice as true."

"Why don't *you* do something about it, champ? You're the bouncer!"

"Because I don't wanna go near 'em. It's gross."

"That's exactly why *I* don't want to do it. And, I'm a little indisposed here at the moment. Can't you call the police or something? Or use the fire extinguisher?"

"Wait," Josh said, raising one hand. "He's coming out. He's looking around. Aaaand he's…he's coming this way."

Elizabeth started putting her dress back on, and I put myself back into my pants while gnashing my teeth. A blur the color of Charles' shirt passed in front of one of the windows and I heard the door handle pulled and released with a clunk.

"Open up." There was something restrained in his voice.

"Yeah, just a minute."

"Open up *now* and let's get the fuck outta here."

"Hey, watch it—" I started to say, but Elizabeth cut me off with a warning look.

"We're going," she said. "Just give us a sec to find the keys so we can open the door, okay honey?"

"Can't you just pull the thing up and unlock it?"

"No, it's broken and we need the electric thingy."

She got her dress arranged and looked at me to see if I was ready for the public. I was, but I wasn't budging, just giving her this sullen look. She did a little shrug with her mouth open, like "Hey, what can ya do?"

I shook my head and climbed into the front seat, hitting the button that unlocked the car doors so Charles could climb in. He looked like he didn't know which end was up. Didn't know what to think of himself and didn't want to face it.

"Just drive," he said. "Just take me anywhere but here."

I drove them back across the bridge to Agnes's house so they could get up early and go to the airport, where Elizabeth gave me her email and told me to write to her and ask if she made it back home safely. Charles went straight into the house and I never saw him again, and Elizabeth gave me a little kiss and said goodnight.

I found out that weekend, from somebody, that Curtis had gotten drunk and committed suicide later that night. He'd been found dead in his apartment by the maintenance guy, who was answering a complaint about the radio being on way too loud since three-something in the middle of the night, and continuing until five the next afternoon, when he went in and shut it off. Curtis was in the bathtub with all his clothes on and his wrists cut.

I try not to think about it too much.

III

So I'm standing there with Tiffany and she's like, telling this story or whatever, and it's kind of, I dunno, annoying the hell outta me that she talks in this goddamned speech pattern and shit, or whatever. It makes me kind of, like, want to kill her. While she's telling this story about some chick who totally dissed her in the jewelry store because she walked in looking like she couldn't afford diamonds, she's striking a pose and vogueing to fill in the blanks left by her limited vocabulary, and I'm trying to remember what her appeal was. How did I even get it up to sleep with her?

Oh yeah, she's hot. But besides that?

"And I was like," she made a look of defiance. "And she was like," she struck her pose of standoffishness, and made a look like she was peering over the rim of glasses she wasn't wearing. "Yunno?"

"Uh-huh," I said, tilting back my head and draining the second half of my Cuba. Yunno, so the next words out of my mouth could be "Oh look, my glass is empty. Excuse me a sec while I go refill it."

"And I mean, like, the *nerve* of this chick, yunno? Dude, it was a Tiffany's. And *my name* is Tiffany!"

"I know. You had more of a right to be there than anybody. Oh look, my—"

"And yunno what the *worst* part was?"

"No, Tiff. What was the worst part?"

And just as her luscious, glossy, strawberry-smelling lips parted to tell me something stupid, my phone rang.

Oh, sweet Jesus, hallelujah! I love you, phone, and I love whomever it is that's calling, I don't care if it's the collection agency again.

I let it ring twice, and the third time Tiffany said "Aren't you gonna answer that?"

I sighed, all world-weary and shit, or whatever, and said

"Yeah, all right, 'scuse me a sec." It took all of my self-control not to do a little dance as I fished my phone out of my pocket and looked at the screen. Hmm, that's an awful long number, I thunk.

I answered it, and this strange half-Spanish, half-upper class English voice said "Hello, Johnny? It's Elizabeth."

I held out a hand at Tiffany and made an apologetic face and she nodded.

"Hi, how you doing? You sound different," I said, walking away.

"Oh, must be because I've been speaking nothing but Spanish all week. So, what's up? I haven't heard from you."

"Sorry, the maid threw away the piece of paper you wrote your email on before I could write to you."

Shit. I completely forgot about writing to her.

"No problem. I've been really super busy since I got back."

"You're calling from all the way over there? And how'd you get my number?"

"I asked my Grandmother and she asked your Mom. And yeah, I'm calling from here. It's no big deal, I'm calling you on Mary's phone."

"Mary?"

"The other owner of the spa, and my room-mate. She's in Korea visiting her husband right now and I'm here in this big house, all alone."

"Really." My eyes narrowed to slits like the Big Bad Wolf's do in the cartoon.

"Yesss. So, I looked you up on MySpace and saw all your art. My friends are impressed. And Mary wants to buy your ballerina painting if you still have it."

My eyes popped open.

"Does she now?" Woo-hoo! Finally! That's the painting *every*body says they love and want, but no one ever buys, and *finally!* Someone wants the damn thing! "I can go by the gallery tomorrow and see if it's still in there." The gallery in the spare bedroom of my apartment.

"I hope it is. And I hear my grandmother wants you to paint a portrait of me."

"She did say something like that." I coulda kicked myself. If I wasn't wearing a pair of my nicer pants.

"Well, you remember how I said I was going to be coming back in May? Well, I changed my mind. I'll be there in just a few days. With*out* Charlie."

And, like a CD skipping, I froze.

"...Hello?" she asked.

I cleared my throat. "Yeah, yeah, I heard you. Somebody was just saying Hi to me so I had to say Hi back. Soooo, a few days, huh?"

"Yeah. And I've had you on my mind pretty much all day, every day, and I have to get laid immediately, so I'll need you to pick me up at the train station and take me somewhere before you drop me off at my grandmother's house."

Oh. Um, okay.

"You're taking a train from Guatemala?"

"No, Einstein, I'm taking a train from the Miami Airport. I'll let you know what time I need you to be there as soon as I've got my itinerary, okay?"

"Gotcha."

"And Johnny?"

"Uh-huh?"

"Write me a nice long email as soon as you get home."

A few hours later, I knelt on the bed of this girl Amanda I'd met maybe twenty minutes before closing time. It was a whirlwind romance; I'd been dancing with Nicole and all them when she approached me on behalf of her homely overweight friend and I told her I already had my heart set on somebody else. She said "Really? Who?"

"I'll give you a hint," I said. "She's within five feet of me." I dunno why this worked. Maybe it was the music and how drunk we all were, and the packed-in closeness of our dancing.

She looked around and then nodded sagely.

"Lindsey," she said with certainty.

"Nope."

"*No?*" Her eyes went wide with genuine surprise. Understandable, because Lindsey was cute as a button, but I had it on good authority she was a drama queen. Amanda looked around again.

"Michelle?" I shook my head. "Nicole." I frowned with narrowed eyes and a head tilt of disbelief that she'd even suggest such a thing. Nicole had been with Rene for like, three years, and Rene was someone with whom you did not fuck.

She looked around again, at a loss. Then I could see it dawn on her. It took a bit to sink in, and when she asked, it was only a formality. "Is it me?"

I leaned in and kissed her. I could feel the weight of her homely overweight friend's stare as I kissed Amanda with that take-your-breath-away kiss my second great love taught me, and we kissed steadily for the next thirty-something minutes until Josh insisted we really had to leave.

We kissed some more outside and long after her friends had said "For real, last chance, we're *going*," for the third time, and finally left.

Then, fast-forward, she was on her knees with her face pushed into the pillows and I caught a glimpse of myself in a mirror out of the corner of my eye. It was the reflection of a reflection in a second mirror, and I looked in surprise, and watched myself for a moment. I had not realized I was that muscular.

There's only so much progress you can see in a mirrored wall at the gym before it looks like nothing is changing, and then you catch sight of your own back in the mirror on the wall of a stranger's bedroom and wow, there are muscles rolling on a beautiful body you didn't even realize you had, and as you watch yourself fuck someone (because it certainly wasn't making love) it hits you.

If this is what everyone wants, if this is what all the other people are trying to get—a whirlwind one night stand with a cute stranger on a Saturday night, and be drunk and still good-looking and know there aren't any strings attached—if this is the pinnacle of everything we strive for, why'd it all feel so empty?

I looked down, at the narrow hips that rocked savagely into mine, and thought, *Jesus,* I'm masturbating, but using another person instead of my hand.

I reflected for a moment on the testimony of that guy who used to play the piano on Hugh Hefner's yacht. He had access to more than enough money, the best parties, the finest top-shelf liquor, the finest drugs, and what many argue are the most beautiful women on the planet. He attempted suicide.

I reflected for a moment on the burnt-out rock-n-rollers I read about who, to the shock and horror of their fans, decide to stop wallowing in glycerine-slippery flesh pits with their band members and groupies and, out of the blue, convert to Christianity and start singing songs about God.

I reflected for a moment on the great night I had, and the great sex I was right in the middle of having with someone whose acquaintance I'd only just made and whose name I was not obliged to remember, and how this was all anybody really wanted in the world, the end-all be-all, and wondered why I was wishing desperately for something—anything—else.

And that's the last thing I remember about the night before the first day of the rest of my life.

Elizabeth and I started emailing each other, writing long messages, and it was nice to read something more than a quick note perfunctorily tagged onto a forwarded video. Her writing style was lucid and so full of wit and innuendo that I really looked forward to checking my email every day. We were telling each other more about ourselves, discussing this and that. She had the most articulate sense of humor I'd heard in a very long time, and I was becoming more and more impressed with her. There was so much more to her than so many of my exes.

She told me about the day spa she owned and the house she shared with Mary that was *the* party house in Antigua. She told me about this guy Michael that had been her mentor in Arizona, and the year she spent living in the Biosphere, whatever that is, and how they were building the Biosphere II, but she'd already taken Mike's advice and gone traveling, ending up in Antigua where she chose to stay and become somebody, and couldn't be a part of it.

She told me about this religious cult she had infiltrated, or thought she'd infiltrated, until she realized she had actually *joined* and believed in it, and was in deep shit if she couldn't unbrainwash herself quick, fast, and in a hurry. She tried to leave, then tried to escape, and then had to be rescued, and she'd tell me all about it some day. Later, though.

I felt something strange in my stomach from then on until she arrived, like that fear I felt every morning in my Freshman year of high school after Lisa Marulli started sitting next to me in home room. Gwyneth Paltrow said in a movie that it was like there was a riot in the heart and nothing to be done, come ruin or rapture. I think that was pushing it, but, well, maybe not.

I opened negotiations with Agnes about the price of the portrait and eventually we agreed on five hundred even, and she wanted Elizabeth to wear the wedding dress from her first marriage. I thought that a little odd, but the old woman grinned wickedly behind her owl glasses and said "Wait´ll you see it."

I got this email today and had to read it twice, and then a third time just for the hell of it. She had a sex dream about me. I'll quote it verbatim:

We were in this wonderful idyllic cabin, kind of a mountains-on-the-seashore type of place, warm, but not sticky. We were vacationing separately and met on beach...we were both with other people whom we promptly ditched and then went back to the cabin together. Strangely, the cabin had both our stuff in it, and after we ditched our dates, the dream turned instantly into a planned vacation we were taking together.

I was naked, and you were in a sarong. I was hippied out, unshaven legs, wisps of underarm hair, trimmed but not waxed. But I felt totally comfortable and secure, like I knew you wanted me that way. Think old drawings from the Joy of Sex.

We talked, and drank some nice sparkling wine, but not much. We didn't want to be drunk for what we had planned next. you brought out a small jar of this amazing honey, it was scented with herbs and smelled like a beautiful perfume, but musky. It wasn't sticky at all, it ran like syrup on my warm skin and you were licking it off of me. up and down my lower legs, my forearms, my shoulders. I tried to get up and give you the same treatment, but you said very firmly "I am going to do what I want, when I want, and if I want you to suck my dick you'll certainly know love" (Author's note: I would never, *ever* say that.)

Then we flashed forward, you were naked and we were on this huge white bed together, very romantic, mosquito netting, candles, exotic flowers, all that. You were naked and tan (no tan lines) and your body was so firm and buff, but you had just a tiny soft hint of a belly. Perfect. Your hair was blonde, and you didn't have glasses. You put a few pillows underneath the small of my back, so my hips were lifted up towards you.

You knew me in the dream, knew how I usually like it. So, you knew that it is easier for me to cum if I apply pressure with my thighs, like pushing my legs into yours or onto the bed. That is why you had the pillows under me, you put me in a position where I couldn't press my legs against you....because you wanted to tease me and didn't want to risk spoiling the moment with a premature orgasm. You ran your tongue up and down my lips, the inside of my thighs, and I could feel your hot breath on my clit but you wouldn't touch it. It felt so swollen, it was like my clitoris was the only thing in room that mattered. You dripped cold wine on my lower belly and let it run all the way down to my ass. I shivered, and you all of a sudden took all of me into your mouth, and I almost screamed with pleasure.

Then, you went back to teasing me....with a cough drop. A red one. you had it in your mouth and would pass it over me as you pleasured me with your tongue. Then, you sat up in between my spread thighs and smiled at me. Your eyes were so full of love, and you were so happy to be there in that moment, and so was I. You told me how long you had craved me, and I told you how much I wanted you, and although we had been kissing, you kissed me in the way only someone who is truly in love can kiss. With complete abandon. It felt like you were really opening yourself to me, that you had never let anyone close enough to kiss you like that. I kept my eyes slightly squinted so I could see you, and your eyes went wide, almost in surprise. You pulled back and felt dizzy. I told you it was okay, that you didn't have to keep going if you didn't want to. your eyes filled with tears and you said you didn't ever want to be anywhere else.

You still had the cough drop, and I was still laying back with my hips in the air. You wet it in your mouth and began to rub me with it, You would intensify the pressure, and back off, over and over. You said you wanted to make me cum like I never had before. I didn't feel frustrated by your teasing, we were in perfect sync. I wanted exactly what you wanted to give me, exactly how you give it. I know we have both had amazing sex in our lives, but this was different. Like, spiritual or tantric or something. Fucking amazing.

you would tease my clit until I was close to coming, but not so close that you would risk ruining it, and then run your tongue down inside of me, and slide your torso and chest across my groin and belly, and wet my nipples with my juices. Then you looked at me and said, "each time I put your beautiful breasts in my mouth, it is like biting into the most delicious piece of ripe fruit like a cherry popping in my mouth." (Author's note: I would never, ever say that, either.)

At this point, I started to sort of wake up, and I was sooo wet, and I think I may have actually started touching myself thinking of you. I came, I think at the same time in real life and the dream.

Then we laid together in that beautiful, comfortable white bed, and slept in the half-awake, dreamy way the new lovers do. It was implied that we would be vacationing for quite a while, and we rested together, knowing we had many more days together to spend as we wanted.

So, that was my dream. I hope you're not freaked out. You owe me a story about what you think of me when you need a little something extra to climax.

   Josh noticed I was drinking less and talking more, smiling a bit more easily, and asked what her name was.
   "Remember that chick from last Wednesday?" We were

outside in the back of Respect's, watching the people who didn't have the courage to dance badly watch the others who did. Boy, Goths can't dance for shit.

"The one with a boyfriend who wasn't her boyfriend?" He lit a cigarette.

"The very one."

"I thought you said she was going back to Mexico."

"Guatemala."

"Whatever," he said, blowing smoke in my face.

"No, it isn't whatever. They are entirely different countries, and lumping them together is like—"

"Whoa there, Cochise. Since when you give a shit about third-world countries?" He gave me a mock-shrewd look and knocked my shoulder. "I know someone who's in lo-ove."

"Shaddap."

"Lovey lovey lo-ove!"

"Fuck off."

"Hey, congratulations, bro. It couldn'ta happened to a better guy. Speaking of that, saw your sex tape today. Online."

"Ha ha."

"For real."

"What?"

"You looked pretty out of it. Probably happened during a blackout."

"Waitaminute. You're *serious?*"

He nodded with his eyes wide at the seriousness of it while taking a drag. I watched the bright cherry of his cigarette as it flared, instead of looking at his eyes. The ground underneath me was feeling a little unsteady.

"With, uh…with whom was I? Yunno…"

"Fucking?"

"Yeah."

"Dunno. Lou showed it to me, said Whatshername forwarded the link to him."

"Oh yeah. Her."

"You gotta be careful, bro. That kinda thing, when you try

to run for office...it'll come back to haunt you."

"Yeah," I said, suddenly feeling *very* naked. "No doubt."

I didn't hit on anybody that night, not even the sure things or the new girls. Not that night or the next, or the next, and when Elizabeth finally called me saying she was leaving Miami on the TriRail, I had been celibate for a week.

She was gorgeous. I could'a sworn she'd only been attractive in the once-you-get-to-know-her sense, but now that I could see her better in the light of day, she had this pearly kind of ultra-fair skin that I had used to hate and found adorable now, and her charmingly crooked teeth was a pleasant change from those perfect smiles almost everybody had. None of that cookie-cutter South Florida beauty that I'd grown so tired of, and none of the punk rocker and Goth hipster slut scene that was so affected up at Respect's. I couldn't understand how I hadn't thought she was the most beautiful thing I'd ever seen.

The word that came to mind when she stepped off of the train was Classy.

She had on a pewter knee-length skirt, matching heels, and a wine-red blouse with wide-brimmed hat, and I thought Nice, a *skirt*. And a hat that isn't just a baseball cap or one of those sporty little K-Fed hats everybody thinks they have to wear, boys and girls alike.

She handed me her suitcase and presented her cheek for a kiss. Also a nice touch, I thought. Not taking things up from where we left off. Backing it up a little. We went to the car and she allowed me to open the door for her without showing that she expected me to, which I appreciated. Some girls will speed up the last few steps and wait at a door, to make sure I and everyone nearby know that she expects to be treated like a lady without acting like one, which irritates me to such an extent that, instead of opening the door, I'll pretend to get distracted by something and make her wait. But Elizabeth didn't do that. She acted like an actual lady.

She told me all about her trip, and what was going on back

home, and the crazy day she'd had right before leaving, and then suddenly put me on the spot by asking me where we were going. My mouth was instantly dry because I'd been driving us to my place. Where we could make love or have sex, or whatever, just like she had said she wanted to do before I took her to her grandmother's house. *Right?*

I was on the hot seat. I knew instinctively that if I didn't say the right thing, and say it with real aplomb, then the whole affair would take a nose dive before it even started. It was one of those things girls do, sometimes without even knowing why they are doing it, on the spur of the moment to test a guy and see if they are considered just an object or if the guy actually respects them. And it shoots them in the foot, like it could right now, her coming all this way for exactly *this* and deciding she might not want it if that's all there was, and ruining a good thing before it happens, thereby wasting the entire trip and being stuck here for a week or two in a relationship that went sour in one fell swoop that *she* caused. But they'll do it. They'll do it on a whim, just outta the blue, because something inside them is hard-wired by millions of years of evolution to make sure the guy they might get knocked up by is the right one, and they just have to test his mettle without warning because dammit, they gotta know.

I bought myself a moment by shutting my mouth and raising my eyebrows inquisitively, asking "Hmmm?"

She was clearly enjoying this, knowing everything that was going on inside my head and hoping that I wouldn't spoil it by acting like a regular, big dumb Male. "Where. Are. You. *Taking* us? At this moment. Because you seem to be driving somewhere specific. Where?"

Counter-intuitive response, I thought. Gotta be a counter-intuitive response. The idiot would automatically say he was taking her to his place to have sex, because that's exactly what she'd said she wanted to do. Didn't she? I mean, am I wrong? And the wimp would hastily act chaste and righteous and eager to please, saying he had intended from the start to take her straight to Granny's house and he'd be back to pick her up at 8pm sharp for a nice dinner at Dempsey's.

And she was a smart one, so she'd probably expect me to know I shouldn't say either of those things, and wanted to see how inventive I could be at a moment's notice. And I'd be graded on it, without forgiveness, whatever I came up with. So I said everything I wasn't supposed to, only with aplomb.

"Well, I figured you were pretty tired from your trip, so we're going to drop by my place and get a few in before I drop you off at your granny's."

She looked at me in feigned shock, which was a good sign, I guessed.

"I'm sorry, what? 'Get a few in?' A few what?"

"Yunno, a few quick ones. Just because I bet you're too tired for a whole, you know. *Love-making*. See? I'm being sensitive to your needs. So, a few quickies to unwind after the trip."

She did that kind of half-laugh, where only one comes out. "Huh. Really?"

"No, not really. We'll pop over to Josh's brother's place to get something to burn later when you put your feet up."

"Oh?"

"Yeah. Why, where'd you think we were going?"

She slowly twitched a corner of her mouth to a half-smile.

I want to get a camcorder and hide it somewhere in my bedroom, then get Elizabeth high and videotape her. The tirades and theories and witticisms that she spews in her verbal diarrhea when she's stoned would make us a fortune, and land us book deals and a running spot on the lecture circuit for years. The first time I got a private screening of The Elizabeth Morrow Show was a few hours after picking her up from the train station, after getting a dime bag, heading back to my secret lair, and firing up the bowl that had gathered cobwebs since my college days.

With *Jar of Flies* on the stereo, we settled in nicely for some foreplay weed, and as soon as the bowl was cashed, she started babbling. One topic segued into another into another in this stream-of-consciousness rant that was fit—at least in *my* mind, at the time—to publish. First thing was about Paris Hilton. I don't even remember what she said; some criticism that didn't interest me, that segued into:

"Babies are the new chihuahuas."

I blinked at her. Then I shut my eyes tight and shook my head. "What?"

"Babies are the new chihuahuas."

"I, um...I heard you the first time. What do you mean?"

"Remember a few years ago when every li'l Hollywood starlet had to have a purse-dog? A tiny little yappy dog sticking out of her purse? And it'd yap and snarl at every little thing and she'd try to calm it down by petting it on the head? And that only reinforced its belief that incessantly *yapping* and snarling at everyone was a *good* thing to do, because it was constantly rewarded by affection?"

"...Yeah?"

"Well, every starlet *had* to have one. Then there were a few trendsetters that got knocked up, so *aaaall* of them had to get knocked up and photographed all preggie and shit, by the paparazzi, and then the big thing to have if you were an emaciated California twit with concentration camp cheekbones was a big swollen belly. Then just a few months later, the paparazzi were swarming around them all snapping up photos of their crotch fruit, and that was the new big accessory. So, babies are the new chihuahuas."

I was still laughing at "crotch fruit" either an eon or five minutes later, when she started in on the Muslim extremists. It was amazing how articulate she was, given the circumstances, and I had to wonder if she'd rehearsed this just in case she ever had an audience that would understand her. Oh, and she loved the word "zeitgeist" apparently.

"I think the majority of Americans don't feel a significant emotional connection with the iconographic locations that are the most famous in the US. That's why it's sort of stupid to repeatedly target sites like the World Trade Center, at least if your goal is to, quote, destroy America, unquote.

"Any literate person with ten thousand dollars could bring the American economy to a screeching halt, for at least a few weeks and make a *lot* of money shorting markets, if they would simply invest their resources wisely. Imagine if over the course of two weeks, random Walmarts, supermarkets, local sports events and megachurches are bombed. Trucks, suicide bombers, doesn't matter. The point obviously, is to create real terror for the majority of people.

"After 9/11, there were plenty of flag wavers and newfound patriots, but from a psychological perspective it is really easy to disconnect oneself from fear of terrorism directly affecting your own life if you live in places like Tulsa or Des Moines. But if grandmothers, PTA presidents, and blond kids in their little karate uniforms are tragically killed in random, relatively unimportant locales, folks'll start caring. If your dad is blown up in a bar with a bunch of his buddies after work, it'll make a difference.

"For terrorists, overhead costs should be low, because the point isn't to kill that many people, just the right *kind* of people. People who could be you, me, or your grandmother. And constructing a bomb that would deliver moderate destruction in a small package is really not that hard. If you aren't worried about trying to transport those things on planes, or into secure areas, then it's even easier.

"So, now, if one had foreknowledge of these events, a few well-placed short sales, commodities speculations, and currency exchanges could be highly profitable. You've accomplished your goal of striking terror into the heart of the Ugly American, and made a fortune on the blood of little-leaguers."

I think this is what they call a Red Flag.

The fact that she could even *think* of such a thing—not only that, but put so much thought into it—meant she had more than a few screws loose.

But I was about to get laid, so I dismissed it.

Then she started talking about this Mayan pumice thing, a Pre-Colombian artifact that was in somebody's house, and she pinched it while drunk, stuck it in her luggage and successfully smuggled out of the country by accident. It was one of those things where, if caught, she could get the death penalty. Anyway, her brother found it and used it to grind up pills for snorting, when she walked in on him. Telling me how trippy it was that he was obliviously triturating Oxycontins with this thing that belonged in a museum. She laughed her contagious laugh, and I fell in love, despite myself.

She told me about how the house she shared in Antigua with her business partner Mary was *the* party house in Antigua, and how I *must* come and visit, I would have the time of my *life*. And she told me about the lake. *The Lake*. Lake Atitlan, which was in the running for the next Wonder of the World, but there was a little problem with it suddenly being horribly polluted. Then she got all serious.

"You know that sad commercial they used to have, where there is a car full of white kids going down the highway, having a good time, and they throw their trash out the windows, and the litter rolls to the feet of an old Indian gentleman, and a tear rolls down his weathered face? Well, it's a crock of shit. Indians are the dirtiest people you'll ever meet. Antigua is a litterbug's *paradise*. The streets are their bathroom, too. Once a year, for three days, they line up outside Renap, the governmental office where they get their papers organized, and they leave the street *carpeted* with empty junk food bags and plastic bottles, and Styrofoam plates. They have *no* love for the land they claim is 'rightfully theirs.' And neither do the ones in North America. Garbage everywhere. Get this: when I used to be one of the Biospherians, I was listening to these others that had used to be contractors, and they had built houses and given them to the Indians to use, for free. The savages put their livestock in the houses, and they themselves slept outside. You believe that? Of course, the goats and cattle destroyed the houses, and the people that built them could only shake their heads at the waste."

The next ten days passed like that. Us getting high and making love, spending time with the grandmother, going out, meeting my friends, having a great time. Being happy.

On Day Three, I was called over to Agnes's to take photos for the portrait, and was astounded to see the old woman fussing over her granddaughter in this...um, rather revealing net-like dress. It was the dress she had worn at her third wedding, and it had a lot of gold material that was strategically arranged to *not* cover anything that needed to be covered. Her nipples, her cha-cha, everything was *bam!* right there for all the world to see.

And Agnes had worn this to her wedding? I couldn't believe it, but I think I covered it pretty well. Her granddaughter seemed to be a little embarrassed, but not too.

I took the photos, many of them, and we experimented with different poses and different angles, getting a few great ones of her seducing me with her eyes next to this lapis lazuli pyramid on a black walnut desk. It took about an hour or so, and then Elizabeth was back in her street clothes and coming over to the house to get high and make love and do another episode of the Elizabeth Morrow Show.

On Day Five, she told me about Brendan, who was like that song *The Speed of the Sound of Loneliness* as if that was supposed to mean something to me. Brendan had this habit of getting drunk and picking fights that he couldn't win, to see who would man up and knock him off his stool. On the rare occasions that someone would, he'd get up, announce that the guy was a right cunt, and buy him a drink. It reminded me of what Josh had said once about men who try to get themselves killed, often over nothing.

She told me Brendan was at present in China, because he took a death-threat seriously, and all the people at Café No Se had a fundraiser to buy him a plane ticket to anywhere far far away.

"Death threat?" I prompted.

"Yeah, my ex-boyfriend has this annoying habit of scaring off anyone he sees kissing me."

She looked at my eyes to see what they'd tell her, but luckily, I'd been affecting a brand of detachment to cover up for the fact that I was stoned and couldn't muster up much in the way of facial expression. Rather than appear vacant, I'd put on an easy face and left it there. She saw it as me not being impressed, which was fine.

"You probably wouldn't back down from Toto," she said. "Which makes you more of a man, but will also get you killed."

I harrumphed and tried to raise one eyebrow, which I've never been very good at.

"Toto? Your ex-boyfriend was named Toto?"

"It's a nickname. He's really Otto-Erick. A big, mostly Indigenous psychopath who tends to put one of the many guns he has about his person to the head of whoever he hears I've made out with and tells them they have twenty-four hours to get out of Guatemala or they're dead. And he means it. And then he comes and beats me up and tells me he'll forgive me if I come back to him."

I was still processing that (weed makes me *reeeeally* slow) and she seemed to interpret that as me being angry, not at her, but for her. I picked up on her seeming to find that satisfactory, as if I had taken the bait and could now be led wherever she'd wanted to take me. Or was I imagining that?

Christ, I hate weed.

"And you *went out* with this guy?" I asked.

"Well, we fell in together and became petty drug dealers, and slowly built ourselves up to the middle echelon. He's a bit higher now and has been keeping his contacts from me so that he'll always have an edge. Anyway, one night, you know how it is. And we were together for two years after that. We've been on so many adventures, you wouldn't believe.

"One thing that amazed me about him, he would create whole worlds inside his head to deceive me. Like, he wouldn't just lie to me. He would imagine every possible way I could've questioned him and create an answer for it, with an entire history for that answer to make it plausible. And he didn't need a notebook to keep track of it all, or if he did, I certainly never found one when I went through his things to see if he'd stolen anything from me. He maintained these gigantic, epic lies all in his head, constantly tweaking them, revising them, and when I would finally figure it out, the scope of them was inspiring."

Inspiring?

That seemed like the wrong word for her to use, but I kept my mouth shut and just listened.

"Finally, I split up with him because I got tired of him hitting me, and he finds me wherever I'm hiding and gives me an ass-kicking then go gets piss-drunk at No Se. I see that look in your eye, but he's got so many of the police in his pocket, either because they're his cousins, or his customers, or just plain friends, that he can get away with firing his pistols drunkenly into the air in Central Park, in broad daylight, with no regard for where those bullets come down, and has never gone to jail. So you can imagine what'll happen if somebody was to off him. If I could get my hands on his phone, though, I could get his contacts and take over his business, and then find a way to bury him."

She watched me for a long moment. I could tell that this was a test, but I couldn't tell what my proper response should be, impaired as I was. I stalled for a moment, running through a bunch of possibles, the pressure on, until I finally shrugged and just pulled her to me.

"What an interesting life you lead."

But this time while we made love, I was quiet.

IV

If they ever make a Cliff's Notes version of this book, they'll be sure to say that this part is called foreshadowing.

For as long as I can remember, I've been terrified of being taken hostage and strapped to a chair or a bed or a gurney, and tortured, and killed. I had nightmares of it long before I knew it was the kind of thing people did to each other, so I've surmised—on those days that I believe in reincarnation, anyway, which come and go—that it happened to me in my last life, and the horror of it has stayed with me well into this one. That's why I can't watch movies where it happens to some poor schmuck without feeling a desperate helplessness for the next few days.

That's why I have a particular hatred for criminals who have been caught after doing it, and it's also why I'll refuse to ever be taken alive, ever, by someone who, say, pulls a gun on me and says I have to go somewhere with them. When I see a news story about some victim's body being found, I wonder what that person thought would happen to them if they just cooperated with whoever abducted them—they'd be forced to endure tickling and colorful language, then released?

No. They'd be held captive, maybe tortured, and probably killed. And sure enough, they were.

That's why I also prepared myself for its ever happening, and been glad for having done so twice in the past. The first time, I maced the guy who held a knife to my throat and beat him with the telescoping baton I kept on me until he was at least unconscious, and then I ran faster than I think I ever have before. I always regretted not having stayed behind to make sure he was dead, but you can never really know what you'll do in that situation until happens. And you can shout at the guy on TV til you're blue in the face, but he's never going to do what you think you'd do in his position. And to be honest, you probably won't either.

The second time it happened, I suppose I'll have to tell a bit of a background story first.

I used to be engaged, to a stripper, of all things. Hey, I was young and stupid. And rich, let's not forget that. So I had plenty of money that I hadn't really worked for and so didn't appreciate, and she was hot and a nymphomaniac.

And hot, horny chicks tend not to starve.

They tend to find stupid, impressionable young men like, say, me, who believe everything they hear if it's said by red lips that kiss them often, and in various places, and they move into their houses, drive their cars, and take all their money. Well, this might come as a shock to you out there in Audience-land, but that's exactly what happened to me.

My dad tried to tell me she was using me as a meal ticket, but I thought he was jealous of what I had and he wanted to screw it up for me. He just couldn't stand to see me with that foxy lady, and tried to rain on my parade. So, I ran away with her. And this might come as a shock to you as well (because it certainly did to me) but my old man was right.

All those late-night promises and sweet nothings, and the Yes she gave me through tearful eyes when I showed her the ring didn't seem to mean anything after all.

Who'da thunk it?

But hey, what's done is done. And if any of you read my book *Friends Like These* you'll know who that was about. All that about her was true too. Well, a lot of it. Definitely all that stuff about her being a narc and wearing a wire to get her own friends busted for a measly seven hundred dollars a pop.

Really, I'm not kidding. I read her diary right before we broke up. Shame I didn't read it before I lost everything.

Anyhow, I'd ended up in Pearl, Mississippi, of all places. Pearl was where the first of those Columbine-style shootings took place, and I know, they ought to say that Columbine was a Pearl-style shooting, but won't, because Pearl is, and always will be, just Pearl. Christ, it's a dump. And I learned there that growing up a rich suburban white kid is something you cannot hide. You can slum it with trash, and try to wash it off of you by covering it with dirt, but they'll smell it on you. No matter how low you sink, you'll never fit in, and they will always hate you deep inside, and stab you in the back when convenient.

Anyway, in the apartment I had gotten with Ginger, the stripper, I took a stab at being an adult and lived through her bipolar fits of rage, where she'd beat the shit out of me while I tried to be brave and not hit her and try, *try* to talk some sense into her. We had the cops called on us every now and then by the neighbors, and I totally sympathize with them, I do. They shouldn't've had to listen to that. Or to the make-up sex we'd have the next day.

Anyway, that Christmas Eve, we were doing tequila shots, as we were wont to do, and she was all lovey-dovey until the eleventh shot. I had tried to keep count, but hey, we were doing shots—without chasers or training wheels—and I couldn't count straight either. But I knew I had to stop her at ten, or she'd flip out. And she was just about to go down on me, *just about to*, when she decided she needed Just. One. More.

Well, that last one must've been Eleven.

When the cops came, they separated us and the nice one that took me into the bedroom told me, since it was Christmas, he was willing to believe that I'd fallen down the stairs and that's how I got all those bites and scratches on my face and neck. I thanked him and told him that's exactly what had happened. Too much eggnog. Merry Christmas, officer.

He went back into the other room, and came right back saying "Sorry, champ. She just assaulted my partner, so she's coming with us."

And I spent that night alone, drinking the rest of the tequila, playing Duke Nukem, and realizing Santa wasn't going to show up that year.

So, while she was in County, she befriended a girl named Tracy. Tracy had gotten herself mixed up with a bad guy in much the same way that I'd gotten mixed up with Ginger, and she had left Las Vegas with him and his partner in crime, this epic idiot named Karen, to drive cross-country in a U-haul full of automatic weapons for delivery in Pennsylvania. They'd been stopped on the highway passing through Rankin County for weaving. The arresting officer had smelled weed, cuffed the three of them, and found the crates of arms in the back of the truck.

Ray, the guy, was a tall and rangy badass with a mullet and a goatee, and Karen was blonde and trashy-looking. Tracy was cute, but was only nineteen and more than a little naïve.

So, anyway, Tracy demonstrated her loyalty to Ray by claiming responsibility for the weapons. He swore to her that, with his lesser charges, he'd be able to bond out and would come back for her.

He got himself and Karen bonded out by this guy named Terry Griffin. If you've lived anywhere in Mississippi, and you heard that name, you'd say "Oh, shit." But since you probably don't, you probably wouldn't. I sure didn't. And Ray forgot all about little Tracy locked up in jail.

I first heard about Ray one cold winter afternoon, about two months later. Ginger had been out on bail for a few days, and she had asked me to lend her the car. She told me to trust her, and if she hadn't, I probably wouldn't've been suspicious. But she did, so I was. She begged and cajoled me, and I was so scared of her losing her temper that I gave her the keys, and she also took a bottle of tequila, some weed she had in her little tin box, and the Xanax she'd had me buy her.

I got a call a few hours later, from her, asking me to pretty please, with sugar on top, bail her out of jail. Again.

What she'd done was go to meet this guy Ray, get him out of the house where he'd shacked up with a hot and lonely single-mom schoolteacher named—you'll never believe this—Charlotte Love. I mean, that sounds like bad TV, doesn't it? What're the odds that Charlotte Love is going to be hot and alone, looking for a man in a bar to be a father figure for her two boys, and she meets Ray, a swarthy and swaggering arms smuggler from Vegas while he's out on bail?

That's what I thought. What're the odds?

But that's what happened. He'd busted a move on Charlotte Love, and weaseled his way into her little home, and not only brought Karen, but also four two-bit hoods that he had found somewhere. Marshall was the only one I can remember, because he was the stupidest.

So, Ginger picked Ray up at Charlotte's house and pressured him into drinking tequila, taking some Xanax and smoking a joint with her. Pressured him, you ask? Yeah. She could do that. If you don't believe me, you don't get out as much as you think you do. Read *Friends Like These* and you'll see.

So, then what does she do? The wily Ginger? She drives the fuck up to Rankin County jail's front door, gives Ray a kiss on the cheek, and marches right on in. Tells them at the counter that there is a crazy man outside who's going to come in with guns ablaze, and try and spring his girlfriend.

They came out like mad dogs and crash-tackled him to the concrete, beat the shit out of him, and dragged him back into the jail he'd so recently left.

They arrested her too, for reasons unclear to me.

So, I bailed her out, again. And we went home. And she was going to thank me in the manner that one would expect her to, but at the last moment stopped, put on a serious face and told me that whatever happened between us, no matter what, I should never ever *ever* get mixed up with Terry Griffin.

"Why?"

"Just don't."

So, skip ahead, skip ahead, skip ahead.

I found myself broke, car stolen, evicted from the apartment after coming home from work to find the place empty, stripped of every stick of furniture and every poster on the wall. And the wily Ginger gone. I had nothing. And I had no family anymore—at the time—because after I'd run off to be with Ginger, and had second thoughts, I tried to go home one weekend and my older brother told me to fuck off, I wasn't welcome back. And that's all we need to say about that. I can't blame him, really.

I went to work the next day, at this restaurant that did barbecue and country stuff, and the hostess came to tell me there were some people waiting for me at the door. I asked "Who?" and she said "Some people who want to talk to you."

I went, and it was a guy and two girls, and I recognized the tall, rangy guy with a mullet and goatee from some of the Sunday visits at County when I'd gone to see Ginger. The girls were trash, one blonde and the other brunette and skinny as a rail. Cute, but trashy.

They introduced themselves and we went out into the parking lot for a cigarette, where they asked me if I had any idea of where they could find Ginger. I told them if I knew, they'd have to get in line behind me, and explained my situation. Ray shook my hand, telling me that his enemy's enemy is his friend, and that I was welcome to come stay with them where they were squatting.

Outta the frying pan and into the fire. Happy for a place to sleep and shower, I went into that den of thieves and further adventure. After work, they took me to the house, introduced me around, and I sized up the situation.

The Real World, but with criminals.

Pulling Charlotte aside, I apologized for being part of the problem she had in her house, yet another verminous bum crashing on her floor, and promised I'd be out of her hair as soon as possible, and that I'd contribute in any way that I could to the household.

We then got high and drank a bunch of beer, and plotted our revenge.

A mob lawyer I had met (don't ask—a lot happened during that whole skip-ahead part) told me what to do about Ginger. He said that, since I had a checkbook with both my name and hers on it, and no money in the account, I ought to write as many bad checks as I could and then get outta Dodge. People would come looking for me, and when they didn't find me, they'd come looking for her. Also, it was the practice of a lot of grocery stores that checks could be written for twenty-five dollars over the total of a purchase, and that money would be given in cash. I went to the four different supermarkets in the area and filled up carts with food for the house, beer for my fellow squatters, cartons of cigarettes, and got my twenty-five dollars. The trunk of Charlotte's car barely closed when we drove home, and I had a hundred bucks for whatever.

We did that a few days in a row, and got away with it, by the grace of God.

Thanks again, God, by the way. And sorry.

Then, one night, Karen took me to this house, where a skinny guy named Scott seemed to trust me immediately, and after hearing that I was the guy Ginger had used up and spat out, he offered to take me with him on a drug deal to earn a little extra. He said he'd give me a gun and, when we made the exchange, we would double-cross the other guys and kill them, keeping the drugs and the money for ourselves.

Uh, gee…thanks. But, um…no, thanks.

Then his sister, Shadow, came through the door with a baby in her arms and told us about the shitty day she'd had at work. From what I gleaned, she worked at a strip club as well. These were upstanding citizens I was running with. Anyway, some guy kept pressuring her to charge less than ten bucks for a blowjob. She finally got him to cough up the money, the cheap bastard.

Yeah, you won, I thought.

She nursed the baby for a bit, and Scott kept staring at me until their father came home.

Now, this big ol' guy was friendly and all, with his long hair, beard, and pajama pants and t-shirt, but there was something about him that made me afraid. I mean genuinely *afraid*. It wasn't just a creepy aura about him. He emanated this deep confidence in his ability to commit casual violence and do quick and easy corpse-disposal at a moment's notice.

I was sitting there trying to be invisible, listening to him talk about how he had gotten another one, and the others brightened up and leaned forward a little, ready to hear a good story. He told them he was at Shitkickers, this bar out the way there, doing shots with Duke and them, and there was this migrant worker came in to shoot some pool and have a beer. Dark little fella, looked Mayan, or something. So he picked a fight with him, calling him all kinds of names and shoving him around in front of everybody, and the guy seemed scared to be the only dark guy in there and the target of all this unprovoked abuse, but showed he had a backbone.

When he was invited to step outside and settle this, he accepted. The entire bar emptied out into the parking lot to watch the spectacle. There, in front of all these people and God and the stars and crickets, this guy took his Glock out of the waistband of his pajama bottoms and stuck it in the face of the poor migrant worker. Tole 'im if he dint git on 'is knees that minute, his brains'd be all over the winda of the pickup truck behind him n' there wasn't nothin' nobody'd do about it.

The Mayan looked at all the people watching, not seeing a trace of pity in anyone's eyes, and, shaking, got down on his knees. He then had it explained to him what would happen if he didn't pull down the front of them pajama bottoms and suck off what he found there.

By this time, the storyteller was looking right at me, and so was everybody else. Poor li'l non-invisible me.

"And I tell ya, boy," he said. "A man can think all kinds a things about himself, how tough he is, and what he'll do, and what he'd rather die before doing, but when it comes right down to it, you stick a gun in his ear and make him *know* you'll pull the trigger, he'll realize that suckin' on a man's dick ain't that bad. Maybe he'll even find out he likes it."

I did that face where you raise your eyebrows and turn the corners of your mouth down, nodding slowly as if thinking it over and going Hmm. Because, really, what the fuck else was I supposed to do?

Then, Karen decided to introduce me to him.

Thanks, Karen.

"Yunno who this one is? He's the boy that hoe Ginger took for all he was worth and left him hanging out to dry."

"Really?" this guy said. "I shoulda figgered cuz I seen 'er in that car I *know* she dint buy, and knew there was some poor little rich boy off crying somewhere. That'd be you, then. Well, howdy do? I'm Terry Griffin."

I shook his hand and felt very small.

"I'm Johnny Yen."

"What, like the song?"

I smiled weakly and nodded.

"Well, Johnny, how you gonna get back at her?"

My mouth was so dry I couldn't talk.

"Smart man," he said. "Keepin his mouth shut."

He then explained that he was the bail bondsman bountyhunter that had gotten Karen and Ray out of County, Ray the both times, and was making sure they didn't try anything funny while they were out. Ray'd already had it explained to him, and Karen corroborated this, that the minute he even thought about jumping bond and running, Terry would smell it and make him perform a homosexual act upon his person. And that badass sumbitch Ray was apparently believing it, and not killing Terry Griffin on general principle just for mentioning it, like one would think. Like I was thinking I'd do. If I had a gun. Which was something I was very terribly conscious of at that moment, the fact that I did not have a gun.

What I did have was an extendable baton and bear pepper spray, which felt so heavy in my pockets that my hands, attached as they were to arms that had wet noodles for bones, wouldn't be able to get them out of my jeans in time to use them. And even if I had used them, they would have no effect whatsoever, except to enrage this evil monster that now towered over me, cornering me helpless in a plastic kitchen chair.

This was the guy I'd heard about, whose three children sat in that room with me. That's right, three. And Karen's not the third one. That baby over there, that Shadow the stripper was suckling, Terry Griffin had that baby with his daughter. And bragged about it. Scott, too, was his catamite son. The three of them all got down together from time to time, and when that baby was old enough, who knows?

I put myself somewhere far away in my mind. I went to Italy, where I had been with my family for the first time the summer before I met Ginger, and ruined my life. I put myself in the Piazza San Marco with all of the tourists letting pigeons land on them. And outside the Palacio Londra near the monument of King Victor Emmanuel, making out with Chantal, that pretty German girl that I'd locked eyes with earlier. I put myself in that antiques shop in Taormina, Sicily, with a drop-dead gorgeous girl who kept interrupting my clumsy Italian to tell me she spoke English, and I insisted that I needed to practice as an excuse to not speak to her. And my dad asked me later why I didn't ask her out, since it was so obvious she felt the same chemistry I did and wanted me to sweep her off her feet. I put myself in that discotheque I'd ended up in later, after I'd gotten up the courage to go back and find her, and couldn't. I went to that antiques shop, which was closed, and went around every night spot looking for her, ending up kissing Maritza, the drop-dead gorgeous bartender while she was on break.

Falling in love with two girls in one day, and then again a week later in Venice with my waitress. And making out with Chantal instead, because my waitress was taken, and Chantal was on vacation.

"Y'all, that there's the look of a man on a *mission,*" Terry Griffin said, startling me out of my reverie. Everybody was looking at me again, and I didn't know why. Griffin was smiling and nodding. "Yep, you see that look in his eye? That boy wants revenge. I like that."

I was glad he mistook whatever he saw on my face for something tough and dangerous.

But that's not what I'm talking about, with the relief that I was armed. I'm getting there, though.

Later that night, we went to this trailer park and met Nicole, this tiny little thing that was cute as a button, except for her being in need of a wash. She was fun and bubbly and she made everyone smile. I found out she was Karen's girlfriend, but she went both ways, and was laughing about insisting that she *didn't* to this one guy, her mostly harmless stalker named —"I shit you not," she told me—Mike Hunt.

"I swear," she said. "His folks named him Michael Hunt and he goes by Mike, and doesn't think anything of it. I mean, can you believe that? So, anyway, he's sweet the way he's always trying to woo me in his gentle-hearted way, but I'm just not into him, yunno? And so I try to make him think that I'm with Karen and I'm just not into guys at all, so it's nothing personal, yunno? But he just ain't buying it."

So, anyway, we score some weed off of one of her trailermates, and smoke it there (which makes no sense to me, because if the dealer just sold it to us, why the hell does he get to smoke it with us? That's like your waiter pulling up a chair and eating half of your steak) and then head off to this place, the Depot, in Jackson. *Jack*son, Missis*sippeh!*

While we're there, we score some acid, and even though I swore in my college days that I'd never do acid again, my life was pretty much in the toilet, and anything that would get my mind off of that fact was welcome. So, I dropped acid.

I don't recommend that. Not in an unfamiliar place with people you don't like, and lots of other people you don't know crowding around you, pushing and shoving. But I was committed to it for between eight and twelve hours, so I tried to make the most of it. We danced and smoked several packs of cigarettes.

At some point, Nicky came up and tugged us on our elbows and shouted at us that Mike Hunt was there looking for her and she needed to hide, and would we please tell him that she wasn't here and send him away? Then she was gone.

So, we went toward the entrance and intercepted a stocky but handsome—in a slow kind of way—guy that Karen introduced as My Cunt. We all tried our damnedest not to laugh, but Christ, we were tripping on acid for God's sake. It wasn't our fault. He didn't take umbrage to us at all, and was perfectly satisfied to hear we were laughing because we were tripping. We laughed a good ten minutes at the absurdity of it.

When our fits had subsided at last, we told him that, sorry, Nicole wasn't there.

"Yeah, but I know she's coming, so can I just wait with you til she gits here?"

Karen was about to say No, and since she was such a shitty liar I knew she'd tip him off that she was, indeed, lying, so I stepped in.

"You know what we ought to do is all wait for her together by the front door so that when she comes in, she'll see us, and she won't get lost for two hours trying to find us in this crowd." A gold star for Johnny, the man with the plan.

That seemed good enough for everyone except Karen, but I shut her up with a look and we started fighting our way to the entrance. I shouted at her when Mike Hunt's back was turned that she should go tell Nicole where we'd be, and she could still have a good time for a little while and then sneak out through another exit. She agreed that it was a good idea, but for some reason didn't do her part of it. She was tripping, after all. And even on a good day, she was still Karen.

So we got to the entrance and plopped down on this ratty yellow couch they had there. It wasn't that big, and we were all squished in together, and we just sat and smoked cigarettes for I don't know how long, when all of a sudden, for no good reason at all, Mike Hunt swiveled around in his seat and laid his legs across my knees. I jumped up and said "Hey!"

Mike Hunt also jumped up and said "Hey!" and bumped into a bouncer, who said something that I couldn't hear. Mike started shouting at him. The bouncer shouted something back. Mike's face got bright red and a vein stood out sharply on his neck, and the bouncer's face got even redder, and I don't know who hit who first, or who grabbed who, rather, because the both of them were on the floor locked in a death grip. The great tide of humanity backed up to give them space, and we watched them roll across the floor toward the front door, then *out* the front door, and down the sloping, railed walkway that earlier had organized people waiting to get in, putting them in one long line. And we crowded the door and watched them go, me thinking "It's amazing how quickly these things can happen" while everybody else was asking "What the fuck?"

Then, I have no idea how, because my back was turned, but the one fight had touched off a chain reaction that led to almost everybody in the club fighting one another. Right there behind us. So, we went outside. From out of nowhere, cops had arrived and told us to get the hell out of their way. I had no problem obeying them, thinking This is not my world.

While digging out more cigarettes and lighting them for us both, I suppose the situation escalated even more, but I had missed it. Now there was a ton of people out on the railed walkway, shouting at the cops. The cops were saying they'd arrest anyone who was still on that walkway in ten seconds, and all the people assembled there were advising them to go fuck themselves. Which, I suppose, was a reasonable-enough response. Yunno, counter-offers and such.

When we were all safely smoking, I noticed Mike Hunt sitting sadly in the back of a squad car parked not too far away. Karen noticed him too, and said "There's Mike! Let's go talk to him!"

A black cop walking past us on his way toward the club said to stay away from the squad car. Not to even go *near* it. No problem, I said. Come on, Karen, let's get Nicky and them and go. We're tripping, I'm armed, and you've got weed in your pocket.

She said Sure, and we turned and started walking toward the road, me saying it would be better if we waited over there in plain view so Marshall and Nicky and them-all would see us when they came out, and we'd be well out of the way of drunk people wanting to hit us for no reason and get us arrested, or cops thinking we were somehow involved in this, and come to think of it, we already *were* involved in it, since it was my knees that Mike Hunt had put his legs on to start this whole mess, when I glanced at Karen and noticed she wasn't there walking beside me anymore.

I turned around and ran back to find her bent over to look in the police car window at Mike, trying to hear what he was saying. What the *fuck?*

"Karen!" I shouted. "Get away from there!"

The black cop came running back, yelling at me as if I was the one with my nose pressed to the glass.

"Didn't I just tell you not to go near there? You wanna git arrested too?"

"Hey, I'm not the one. I'm way over *here*."

"Git away from there now!" he screamed.

Karen made a face of Oh, right, and came back toward the two of us, and the cop told us to stay the fuck away from the cars, looking at me as if I was the one who had ignored him, not her. I had enough of this in the Bahamas. No thanks.

"You got it, sir. We were just leaving," I said.

"You better. Or I'll have your ass in County."

I grabbed Karen by the elbow and walked her toward the main road again, telling her "Jesus Christ, don't do that, we're tripping and you've got weed! Are you outta yer cotton-pickin' *mind?*" when she shook her arm out of my grasp and screamed at me "I am *not* gonna let some motherfuckin' black *nigger* talk to me like that!"

Oh shit.

I threw up my hands helplessly at the thought of us ever getting back to Charlotte Love's house that night, and before I could open my mouth to tell her what an idiot she was, she was crash-tackled to the pavement by the black cop, who had come flying out of nowhere. I watched in shock as he wrestled with her on the ground until my voice came back to me.

"Okay, calm down, she's had enough."

"You want some of this, too?" he snapped.

"No."

"Then shut yo' mouth, cracka. I'll arrest da both y'all." He yanked her arms, one after the other, behind her back.

"What's the charge?" I asked, knowing I shouldn't't've, but it pissed me off that he'd suggest it. I hadn't done anything wrong. Not really. And he'd called me cracker.

"Calling me a nigger."

"But one, I didn't. And two, that's not a crime."

"Oh yes, it is!"

"Really. But you can call me *cracker?* What law says so?"

Ooh, I shouldn't't've said that, either.

He'd just cuffed Karen's wrists together and was trying to drag her to her feet, glaring at me saying *"Cracker,* I can make a charge up and who they goan believe? Me, or yo punk ass?" when the radio on his belt crackled in that drive-thru window voice, telling him something. We stared at each other, me feeling my extendable baton in one pocket of my jeans and my mini-fire-extinguisher-looking can of bear spray in the other. Thinking 'If he goes for his gun, I'm going for the mace and my stick, and Karen's going to get a faceful of that spray, but she deserves it, and I'm not going to jail for her.'

I really didn't want to have to take that step, though. I knew it wouldn't end up well for me. I was tripping, and knew my judgment would be questionable, at best.

The cop hesitated a moment, making sure I felt his eyes burn into me, then dragged Karen back across the parking lot to one of the empty squad cars, throwing her in and running back up into the club to arrest more people.

Nicole came up to me then, asking what the hell'd happened, where was Karen? I gestured with a nod toward the police car she was sitting in, looking forlorn and crying, as if it wasn't her own damn fault.

"Holy Christ," Nicky said. We went over to the squad car, just not as close to it as Karen had gotten to Mike Hunt's. "What're we going to do?"

I shrugged, drawing on my cigarette, and while exhaling a

cloud of smoke said "She's got the weed."

"Yeah, I know. What're we going to do?"

I tried not to laugh, and we left, walking up to the main road and down it in the general direction of Pearl. At some point, we started holding hands.

Later on, back at Charlotte's house, we told the story to Ray and he told us about other times that she'd done stupid-ass things that he'd had to bail her out of. Then we called all of the jails in the greater Jackson area, looking for Karen and wanting to know her bail. All of them told us they didn't have anybody by that name. We figured maybe she just hadn't gotten booked yet, so we waited a while, telling more stories, and tried calling again. Every jail told us they still hadn't heard of her and wanted to know why we were calling, so we hung up.

Then, surprise of all surprises, Karen called us, telling us we needed to come pick her up at such-and-such hospital before TJ-fuckin-Hooker got back.

"*What?*"

"Just hurry up and git here!" Click.

So, with Ray driving Charlotte's car, we found the hospital and asked about Karen, and were told she hadn't been released yet, just wait there in the waiting room, and we tried to, we did, but *man*, we just couldn't. I mean, Christ, we were tripping still. Well, not Ray, but he wasn't with us. He was out in the car, hiding from authority figures. The smart one.

That cold, sterile and unentertaining waiting room was a horrible place, and the receptionist kept looking at us, like she *knew*. If you've ever been on acid, or at least on any drug in a place where you should really *not* be on drugs, you'll know what I mean. We could tell that she knew. The bitch.

But, eventually, Karen came out and told us with a big smile that we needed to git somewheres.

"What happened?" Nicky asked.

"Let's just go first, okay? Let's *go*."

So we went. We got in the car with Ray and she told us all about how she'd faked an asthma attack and managed to work up some kind of froth in her mouth that impressed TJ-fuckin-Hooker enough to drive her to the hospital, drop her off, and tell her to wait there until he got back. Like anyone would. Yeah, I'll just sit tight and wait to go to jail.

So, we got back to the house safe and sound.

Terry Griffin had somehow heard about it, and he came by the next day to lecture all of us as we sat around the dining room table. Apparently, he is all buddy-buddy with many of the cops in the area, what with being a bail bondsman and all. Jesus. I don't doubt you'd think I'm making all this up, but I swear, it's true. All stereotypes aside.

So then, after the lecture was over, a lecture by *this* guy, of all people, this monster, he turned to me and told me he knew where my car was. Gave us the address, directions, everything, and said if I wrote out a writ of repleven and waited forty days I might get it back. Or, I could go steal it back tonight. Out of the goodness of his heart, he told me all this.

Two of those verminous two-bit hoods, Karen, and the skinny chick whose name I couldn't remember, all volunteered enthusiastically to help me. Ray and the smarter of the four two-bit hoods stayed quiet, staying out of it, but Marshall and Brent were all for it. Marshall said he definitely should go because we didn't have a key, and he knew how to hot-wire cars.

It was decided that Karen would drive Charlotte's car, and the other three would accompany me as my accomplices and bodyguards.

Jesus. I was doomed, I thought.

But we went, that night. We followed the map that Terry Griffin, cartographer extraordinaire, drew us, and drove for a lonnnng time out into the boonies. Out where there weren't streetlights for miles around, and the stars spattered the vault of the sky in that way that makes you unable to avoid feeling insignificant. Where you can see Every. Last. One. And know that they are anything but close together, and every one of them is the sun for somebody else, and at least one of the far, far away people for every one of those suns is someone who's trying to get something back from the bitch who stole it, and looking up at their sky and wondering if there was someone out there thinking the same thing. Or maybe it was just me.

Marshall wouldn't shut up about what we were going to do and how great it was that we were about to do it, and blah blah blah blah blah. Brent wanted to spark up a joint, but I told him No in a voice they'd never heard me use before, that I don't think I've ever heard myself use before. I told him with this strange authority that we needed to be calm and have our wits about us. That this wasn't some little field trip we were on. Sheesh, they all said, but obeyed.

And everyone was quiet until we finally came upon the turn. The left-hand turn into the woods and onto an even darker road. We killed the headlights and followed that darker road another two miles, the longest two miles I think I've ever travelled, until we finally spotted lights through the trees up ahead. It was one of the only two houses out there, the home of Ginger's sister's boyfriend, and the one next door was that of the boyfriend's mom.

I told them to pull over and wait there until we came back, and not make a sound. No music on the radio, nothing. Marshall came with me, and we crept toward the lights streaming through the leaves of the bushes until the driveway was in view, and I cursed under my breath. It was gravel. And anybody who's ever walked across a gravel driveway knows those little pebbles were the cheapest burglar alarm you could ever have. You couldn't creep up on a deaf person crossing a gravel driveway.

But there was my car, as I lived and breathed. My candy-apple red convertible that Ginger fuckin Newton drove out of my apartment's parking lot one cold cloudy day, that I thought I'd never see again.

My car.

We placed every step with the greatest care, feeling like kids trying to sneak out of our parents' house and creaking our way down the stairs. The outside lights were on, and we could see the blue light of people watching TV seeping through the windows. I figured that they would be so absorbed in whatever they were watching that they didn't hear anything, and they probably were, but that didn't matter the way those pebbles crunched under our feet.

We were finally beside the car, and I touched it lovingly, feeling so many emotions surge through me that I'd been repressing. The cold metal made me suddenly face how badly I'd fucked up, running away with that whore and ending up broke and homeless in the armpit of America. How every romantic dinner I'd cooked, every bill collector I'd paid off, and every late-night promise we'd made to each other was for nothing. How I was just another dumb guy, used up and spat out like a hundred billion before me, lost among them all like the stars above me, not special or unique in any way, not original at all.

We got to the doors, tried them, and found them open. My heart leaped! The interior lights came on, and I held my breath, praying that no one would see the sudden yellow beacon in the corner of their eye and come out to catch us. We hurriedly prized the doors open and slid in, smelling the leather seats and the stale cigarette smoke, shutting the doors again with a soft chunk, and sat there in the dark for a moment.

When I'd gotten my breathing back down to normal, I asked Marshall if he was going to hot-wire the car, or what?

And then, of all the disastrous things that could've happened, Charlotte's car came rolling past with the headlights on, every face staring at us in the windows. We watched it in shock, watched as it rolled past us, turned into the neighbor's driveway, the boyfriend's mom's driveway, a crunching gravel driveway just like this one, and three-point-turned its way back onto the road to roll past us back the way it'd come. I couldn't believe it.

"Hurry!" I hissed at Marshall.

He stared blankly at the steering wheel.

And, Jesus Christ Almighty, I couldn't believe it, but here came Charlotte's car again, Karen doing *another* fricking drive-by, ever so slowly, all eyes glued to us, and they noisily turned around in the boyfriend's mom's gravel driveway *again!* And came back watching us oh-so-casually as if they were just innocent passers-by.

"What are you doing?" I asked them silently.

They stopped. They actually stopped.

"What?" Karen stage-whispered.

"Get the fuck outta here!" I lip-synched.

She put on a face of Ex*cuse* me and drove away in a huff.

"What's the hold-up, Marshall?" I asked.

And before he could say anything, a light came on in the neighbor's house, a poofy-haired old woman standing in her extra-large Stone Cold Steve Austin night-shirt coming to look through the window at her driveway. We stared in horror as Charlotte's car came back, ever-so-slowly, so painfully slowly, and pulled into the old woman's driveway, and she cupped her hands around her eyes to look through the window instead of at her reflection in the glass to watch. And she watched them three-point-turn their way back onto the road, each of them making a point not to look at her, as if that would make them seem less guilty of something, and they rolled up on us once again. Staring at us. I looked at the old woman again, saw her in the other window, the one facing us now, following the gaze of the people in the car and looking straight at me.

To her, I wasn't a poor boy come to get his car back. I wasn't some victim of a lying, thieving whore. I was somebody in the car in her son's driveway in the middle of the night.

I watched her pick up a phone. I watched her punch in a number. I listened to a phone ring in the house in front of me. I saw a shadow rise in the blue flickering light of a television.

"You don't really know how to hot-wire a car, do you, Marshall?" I asked.

He shook his head.

I opened the door, flooding the inside of the car with yellow light, watching the sister's boyfriend's mom's head snap around to lock eyes with me, and bolted. I heard Marshall coming out and crunching the gravel behind me, and the front door of the house opening, and the heart-stopping cracks of gunfire, and I ran like I'd never run in my life.

Charlotte's car sped off down the road, and somehow I'd caught up with it, was running alongside it, when it slammed on the brakes and was behind me. I was still running. I don't know if I felt the hot wind of bullets passing my head, or if I'd only imagined it, but I was certain of it while replaying it in my mind later that night.

The car picked me up somewhere down the road after they'd gotten Marshall, throwing open the front passenger's door for me, and they were laughing at me for running so fast. I wanted to scream at them for their stupidity, for ruining any chance we'd had, but I had no breath for it. I wheezed and sputtered and regretted every cigarette I'd ever smoked for the next few minutes, listening to their excited bullshit.

We went back the way we had come, and found the main road, hanging a right, and surprise of all surprises, there was a police car waiting for us with his red and blue lights on.

"Oh shit!" Brent and the skinny girl said.

"Oh shit, oh shit, oh shit *oh shit oh shit!*" Marshall started to scream, but I reached around to grab him by his hair, yanking him forward and hissed at him to shut his fucking mouth before I killed him.

"I'll do the talking," Karen said.

"If you talk the way you stayed the fuck put," I started to say, but the cop-car's door opened and out stepped one of those double-Y chromosome mustache types that doesn't fall for anything. This guy was the kind of guy who was born to be a cop and when you see him coming you know the game is up.

He came over to our idling car and blinded us with his flashlight, frisking us with it, and told us to shut our engine off. Karen started to say something and he said Shut Up with the coldest authority I'd ever heard. Karen froze. I reached over and shut off the engine, feeling my extendable baton in one pocket of the same jeans as the other night, and my can of bear mace in the other. Thinking, fuck, not again.

The Cop studied each of our quiet faces, and the beam of his flashlight came back to me. He must've seen something in my eyes that he didn't see in the others'. Something actually going on behind them, probably.

He regarded me for a moment, then spoke.

"Step out of the car, please."

"Officer—" Karen started to say.

"Shut up."

Feeling the need to swallow, but really not wanting to, I obeyed. Opening the door, climbing out onto the pavement, I found myself listening to the crickets and the night-birds and the buzzing flies, the rustling of the leaves all around us, as if this might be the last time I ever heard them. I faced the cop.

"Come over here, son."

I went, feeling all eyes on me. The Cop shone his flashlight on me while we spoke, and I felt naked because of it.

"What's your name?" he asked.

"Johnny Yen. John." My voice sounded very far away. I kept having to shift my weight from one foot to the other.

"Where you from?"

"Um, Palm Beach, in Florida."

"You're a long way from home."

"Yes sir."

"What are you doing in that car with them?"

I watched his Don't-lie-to-me-boy eyes, thinking up and discarding a dozen answers before I finally spoke.

"I'm just trying to get home, sir."

He watched me for another long moment.

I felt those stars above me, and the wind.

"I got a call somebody was breaking into a car," he said, prompting me.

"I don't doubt you got that call from Jim Davis's house, where we just were," I said, surprising him. Doing that counter-intuitive response when he'd expected me to lie to him, tell him Oh no, sir, we just happen to be out here in the middle of nowhere at the same time a few shots are fired at trespassers. Trespassers who've magically disappeared, no less.

"Go on."

"Up until just a few minutes ago, I was engaged to Ginger Newton, Jim's girlfriend's sister."

Something flickered in the Cop's eyes, and he twitched his mustache in the faintest smile.

"Maybe you've heard of her," I said, catching on.

"Oh, I've come across her once or twice."

"Well, then you know what she's like. And I just found out the hard way."

He was nodding, knowing exactly where I was coming from, and sympathizing.

"So, we've been having a little party out here, all of us, and she had one drink too many and started in on me about something some other boyfriend did to her years ago that now she's blaming me for."

"Or her daddy," the Cop said. Whoa. There was a thunderclap of subtext in those three words that spoke volumes. He definitely knew about her.

I nodded, doing a subtle lean in towards him, a lean and nod that said Yeah, yunno what I'm talking about, and we're on the same side here. He warmed to me.

"Go on," he said again, less like a cop this time.

"Well, she laid into me and Jim broke it up, but some of these idiots in the car there started acting up and we all got kicked out. We got the one girl at the party who wasn't drinking to drive us back home, and Ginger was screaming about how she was going to call the cops and have us all locked up. Screaming with spit flying everywhere. Yunno how she gets."

"Boy, do I," he said. "Okay, Mister Yen, why don't you and the kiddies run along and stay out of trouble for the rest of the night. If I were you, I'd get myself a bus ticket home to the beach and forget all about her. You'll be better off."

"Yes sir. Thank you sir."

"You bet. Have yourself a good night."

"We will."

"Hi, Officer Dobbs."

We both stopped and looked at the car, at the gaunt and ugly face of Marshall leaning across the skinny girl to smile out the window at us.

"Hi, Officer Dobbs."

The Cop shined his flashlight into Marshall's face, making him squint, baring all his teeth in a big stupid grin.

"Why do you know my name, boy?"

"You don't remember me, Officer Dobbs?"

"Should I?"

"Ya broke my nose when you arrested me a month ago for possession. How you doing?"

I could have died at that moment.

You. Fucking. Idiot.

Officer Dobbs turned cold again and his hard eyes came back to burn into me. He knew in that moment that everything I'd said had been a lie. It was all over my face and there was nothing I could do to hide it.

We were as home free as we could've been, 'cept without my car, and in one fell swoop Marshall had to open his big frickin' mouth and ruin everything.

In that moment, I realized something a more observant person would've put together the first day walking into Charlotte Love's house. All these wankers, these two-bit hoods and verminous skanks, they'd all met in jail. That's how they came to be with Ray and Karen in Charlotte's house, all of them getting a free ride while they "got back on their feet and got their shit together and blah blah blah." This is what they do, get locked up and do a stint and get out and then get locked up again. Get their three meals a day and a roof over their heads, their laundry done, all at the expense of the county.

In a flash, I saw my future. We'd all go downtown, sit in a bullpen for a few hours awaiting booking, bitching about our bad luck, and then get processed into a cell or a dorm of cells, and say Hi to all the people they know in there and play cards for a bit, and wish we had cigarettes. And maybe I'd get my ass kicked by someone or other. Or raped. All because of the stupid-ass people in that car.

Bull. *Shit.*

So I risked everything and made one desperate stab at my freedom, looking wide-eyed over the Cop's shoulder at Marshall and yelling "No! Don't shoot!"

The Cop wheeled about, yanking his sidearm from its holster while I yanked the bear mace out of my pocket, sliding the safety off with a *snick!* In that split second, the Cop was jumping out of the way to avoid whatever shot might be coming at his back, extending his fist with his pistol in it, and the car full of idiots was looking at us in confusion, and I sprayed all of them with the mace and took off running. I heard gunshots and screaming, the Cop firing blindly, and a moment later, return fire from whoever was packing in the car.

I ran for my life, down the main road farther and farther from Jackson, Mississippi, without knowing where I was going and not caring, either. Running until my lungs burned and my body ached, running until I ran like a girl, until I could hear nothing except the crickets and the night-birds and the wind, and the stars.

I hitch-hiked to Biloxi, where I found a casino and managed to make a little money playing blackjack. I kept it low, making five-dollar bets and letting either God or destiny or just good luck tell me when to ask for another card or stay with what I had. I started with twelve bucks, got up to seventy-five, lost one hand, and quit. Rather than risk losing it all, I took what I had and was happy with it.

Out of curiosity, I went by the one-arm-bandits and had a look, sorely tempted to put a coin in one of them. All those lemons and cherries and flashing lights, and the warped reflections of other machines and lights in the faux-gold surfaces, and the bleeps and whoops, all of them promised me sudden wealth and happiness and a way out of this mess.

Then I saw a middle-aged, frumpy old woman full of loneliness, sidle up to a stool with a bucket full of quarters. Like an automaton, she began feeding coins into the slot and pulling the lever, watching all the lemons and cherries and pots of gold roll blindingly until stopping one after the other, to disappoint her. I watched her do it seven times, each time marveling at her faith, and feeling sorry for her, until the miracle happened.

Chunk! Chunk! Chunk!

Three bunches of fat purple grapes!

The alarm sounded, a metal bell banging away at the top of the machine while lights flashed and quarters began to spill out of the wide mouth and into her lap. She gaped at them in shock, watching them fill her blue gingham lap and overflow, jingling over her legs and onto the carpet beneath her stool.

It only took a few seconds, but those seconds seemed to last a lifetime, and when the last coin came banging out of the machine to fall with a silvery *tink!* she stared in baffled wonder. I realized I was grinning from ear to ear, so happy for this poor woman until she picked up one of the quarters and stuck it into the slot, pulled on the lever, and watched with sad and hopeful eyes.

I couldn't believe it.

I shook my head in disgust and walked out of the casino with my seventy dollars, started thumbing rides to Florida.

Years later, I was in a room when that show Cops was on TV, and I watched with half an eye until a car was pulled over for weaving. The officer ran the tag, went up to the window to talk to the driver, and I heard Karen's voice. After he asked her to step out of the car with her wasted companion, she said proudly "I'ma tell you right now I got warrants for mah arrest in Mississippi."

Apparently, they were in Arizona. The cop asked what her warrants were for, not without a little surprise at her volunteering that information. She seemed to think for a second, and decide to opt for a lesser charge than arms trafficking and jumping bail.

"Bad checks," she said.

Christ. What an idiot.

The officer cuffed her and stuck her in the back of the patrol car, where she started crying about the rotten world and her bad luck. I take some strange consolation in knowing that she didn't get shot or go blind during my escape that night, years ago, but she didn't learn anything, either.

So, that was the second time I was glad for being prepared and carrying a nonlethal weapon in my pocket everywhere I went. I suppose it took longer to explain than I originally intended, but I think it was worth it to say all that, so you, dear reader, would appreciate the circumstances and not think I'm just some bastard who hosed down a cop with bear mace to get out of a trespassing charge, or betrayed a car full of innocent friends.

And so anyway, if Cliff's Notes ever makes their version of this book, they will probably point out that this little story is what they call "foreshadowing".

And, I know it's a long shot, but if you're reading this, Officer Dobbs, I'm sorry.

V

So, with that in context, the whole misadventure with Ginger, and I only just scratched the surface there, you can get an idea of where I'm coming from when I start hesitating over Elizabeth. *Deja vu,* right?

In the next few days, though, I decided that she told me that because she was stoned, and a girl, and wanted either to impress me or make her attractive to my knight-in-shining-armor side with her damsel-in-distress appeal. A bit of a drama queen, maybe, but comparing her with the whole conga line of anonymous or unmemorable lovers passing through my bedroom in that liquor-blurred era, she was a definite improvement. So, I dismissed whatever it was I'd thought about her.

We started talking about the portrait that her grandmother wanted me to paint, what style it was going to be, what colors we would use, and then we started talking about art in general. And this is where I'd like to call a quick time-out.

While I'm on the subject, this is something I've wanted to say for a long time.

I was privileged to attend what's touted as "the most prestigious contemporary art fair in the Western Hemisphere" according to Jan Sjostrom of Palm Beach Daily News. I doubt there's any word in any language strong enough to express the disappointment I felt when I saw the drivel that qualified for display. There was, I shit you not, a very large blank white canvas called (of all things) "Untitled" that proudly took up far more wall space than it should have. Not too far away were two others—a diptych, no less—that were also white but had very small black specks meant to represent people out in a snowy wasteland.

Further on, two desk lamps were clamped to a small shelf and turned on, facing the wall, making what looked like boobies of light.

And I can't leave out the three boxes of Cheerios on another shelf. Three boxes—or boxen as I prefer to call them—of Cheerios. And not even Honey Nut or Apple Cinnamon or Yogurt Burst. Just plain ol' Cheerios. If anyone can claim that a box of Cheerios is art, it should be the guy who designed the box, not some jackass who just went out and bought three of them and glued them to a shelf.

Same applies to the famous Andy Warhol Campbell Soup can that was also hanging. Where is the credit assigned to the mastermind who planned the label of that can?

I saw an even worse plagiarism later on. Some prick did a blatant copy of Vermeer's Girl with a Pearl Earring and made only one tiny alteration: he gave her a lazy eye, making it a grotesque parody, bereft of any artistic merit.

Now, I know...I know...who am I to judge? Who out there has the right to decide what is art and what is not?

I'll say this now and stand by it. If a work can be replicated by vandalism, it isn't art. If somebody else did it first and you copied it, it isn't art. It is theft. If it's a blank canvas, it's an affront to everyone out there who makes the smallest effort at painting, and believe me, there was no small amount of tiny effort shown prominently. Squiggles. And not even *coordinated* squiggles. And certainly no evidence of skill or creativity. Most of this crap is acceptable only when the artists' parents affix it with magnets to refrigerators.

It's about time somebody called Bullshit. This is the Emperor's New Clothes and nothing more, and it is insulting to everyone who views it, because it is basically giving them the Finger, saying "you're all dupes who will accept our contempt of you and love us for it."

Maybe my art isn't Art. Maybe it is merely craft. But I can promise you that every effort of mine is to present the world with something as close as I can come to art, and I can sleep at night. How the gallery reps at Art Basel can is a mystery to me, though.

Abstract artists, free-verse poets, or weirdo musicians who

play discordant notes, all of them were the kids who got trophies for participating, and are now grown up, thinking that all they have to do is throw paint at something, or dash off a couple of words that don't rhyme but "ache with significance" or leave a piano out in the rain for a year and then play it and get the same praise as someone who studies theory and puts in the effort and makes something truly beautiful. And the next schmuck who says that beauty's in the eye of the beholder can sick my duck. That just encourages under-achievement.

The first abstract artists painted their way to give the Finger to someone. Everybody else who copies them does it just for the ability to say they are artists. Ditto with e. e. cummings. He was a revolutionary poet, back when being a poet used to *mean* something. But since so many people've followed in his footsteps, poetry has become diluted to the point that it's largely ignored. A man who dresses up as a woman shouldn't get the right to be called She, just because he would rather be a woman. Poseurs shouldn't get to be called artists or poets or musicians just because they *want* to. You wouldn't eat in a restaurant where some pompous and self-righteous chef charges five hundred dollars for a pine cone in shit sauce, would you?

If a figure skater in the Olympics just threw herself on the ice and slid a few feet on her stomach, should the judges give her anything other than a Zero? Of course not. And it would be an insult to all the other figure skaters who trained hard for the competition. So why should art be any different?

This guy Lance, owner of one of the biggest art licensing companies in the US, says he went to the Tate Museum with his wife to see the new "important" art exhibition of Blah blah blah bullshit, and gave it some credit for being Whatever, but couldn't help but notice that they were alone in the room. They went all through the museum seeing no one else until they got to the Pre-Raphaelite exhibit. That room was *packed*.

Ditto that time I went to Rome and took a tour of the Vatican Museums. The guide said that if we spent one minute in front of every piece of art there, we would be there for over thirteen years, and after the tour, I believed him. We spent a few minutes in most of the rooms, and he had a lot to say about many of the pieces, but when we came to the Modern Art exhibit, he led us swiftly through each room—whoosh, whoosh, whoosh—until we came up to the tail end of another tour and had to wait. I appreciated the time we didn't waste looking at the things we passed, most of which was nonsensical, but what really impressed me was that *not one* of our group (and it was a large group) said 'Hey, hold on! We're missing all the good stuff! Stop a minute and explain this...green...thing.' Not once. In fact, the only comment I heard was the gentleman behind me, when he made his family laugh by saying 'Ah, over here on the right is a fine example of 1970s wallpaper.'

Nobody was interested. But everyone marveled at the realistic art in the rest of the museums.

Now, there are some people that say realistic art was made irrelevant by the advent of the camera. If you want a picture that looks exactly like a photograph, why not just take a photograph instead of going to all the trouble painting a realistic painting? The answer to that is on the faces of everybody who sees a photorealistic painting. I've seen people passing by an art gallery, glance inside and see a photo that'd been printed on canvas, and stop in their tracks. They came in and said Oh my God look at the incredible detail on this painting! Omigod, look at all that *detail!* Who painted this?"

And when they were told it was just a photograph, they stormed out of the place in disgust. That's why realistic art has not been made obsolete by a camera, and never will be. The fact that someone with talent sits down and puts forth great effort and discipline to create something beautiful is what impresses people. Someone who doesn't take art seriously, but wants to be called an artist, is just mocking all true artists and all people that know what they're talking about.

That's all there is to the argument, and the best response to someone saying you're a Nazi or a Philistine, or in the very least, a bigoted reactionary, is just turning a cold shoulder and not wasting your time with them. They have created a kind of elitism where they are "superior" to you without actually being so, and no logic will make them accept that they are not.

Now, the reason I bring this up, Elizabeth and I attended an art exhibition that I didn't tell her I was in. The show consisted of a 'performance,' some of my work, a few nonsensical pieces of garbage (childishly executed portraits of people with maggots coming out of their noses and ears, etc) and the efforts of a girl in one of those military officer dress caps, and a jacket with epaulets who put a bunch of wigs in frames. Seriously, wigs. All kinds, from curly blonde to straight black and every variation, all pressed flat behind glass. They had names like "Perseverance".

Mine were the only ones people were really looking at, and I even sold a few (seven, actually) but the other "artists" kept up their smug and haughty attitudes, and even put my work down as soulless corporate crap. Corporate? Um...

Later, while we were talking about it over drinks, Elizabeth made a few points that led me to write the manifesto for my own new art movement. I published it on various art forums on the internet the next day and let it spark a bit of controversy. For those of you interested in this kind of thing, I have reprinted the rant and its responses (with their spelling errors unchanged) below. For those who are not, you can skip ahead to the last page of this chapter.

The art of today reminds me of the stereotypical American college student, who takes money from his parents with one hand while flipping them off with the other, and here's why. Sir Isaac Newton wrote that if he had seen further than others, it was because he'd stood on the shoulders of Giants. Since that time, we as a species have climbed exponentially higher, to heights undreamed of, and now pat ourselves on the back while we stagnate. We have arrived at this pinnacle of human achievement not by our own works, but by those of our forefathers, yet the methods of those who got us here are disdained as outdated and obsolete.

Since *fin de siecle* and the beginning of the last century, quite a few pompous "manifestos violently rejecting the past" challenged the established values and changed the face of the art world so that people of little or no talent could also be included. Since then we've been inflicted with mountebanks making a pretense of being artists, abetted by charlatans making a pretense of being art critics, all of them playing at a charade claiming that paint splashed on canvases without any regard for form or color or substance signifies some kind of elusive talent. Out of that comes an audience of lemmings arguing the relative merits of different types of drivel and trying to one-up each other in their praise of the Emperor's New Clothes. Anyone capable of free thought who calls this exactly what it is, well, "merely doesn't get it."

These manifestos for defiance of set rules in art, poetry, and literature have been steadily denouncing every guideline there is under the assumption that defiance of rules is somehow new. A generation always disdains the previous one as if the parents are ignorant and the children are plugged into something of which anyone older is clueless.

We claim that times are changing, but examining history shows the same patterns in tedious repetition. What we seem to have forgotten is that those established "rules" came out of millennia of trial and error, that all this experimenting that characterizes the past century was already tried and found wanting, that the reason we are where we are today is because our predecessors *knew what they were doing*. The ground was broken enough to bring us here, and all subsequent groundbreaking is merely a throwback to primitive attempts that failed and should be forgotten. And, since the novelty of being *avant garde* inevitably wears off, we go the way of the jaded. We degenerate and deviate for new ways to achieve the old feeling, and the quest for something fresh has strange horizons.

A common argument supporting this pursuit is Freedom of Expression, but I've found we are so "tolerant" of "diversity" that anybody who expresses a belief not on an approved reading list is instantly condemned for being "intolerant." And that makes sense how? Those who take that argument prevent speech in the name of defending it, and backslide in the name of progress. Thinking for oneself and questioning authority has been perverted into defiance for defiance's sake. It leads to crucifixes in jars of urine which, if Sacrilege isn't allowed to be a condemnation anymore, at least falls under Bad Taste.

It leads to Howard Stern saying "penis" on the air just because he was told he shouldn't. So now we can say "penis" on the air. Now what? Has anything improved? What has been gained? Nothing. Just "blah blah blah penis blah blah blah." It leads to Jackson Pollock and Tracey Emin and Damien Hirst. It leads to garbage that is Art *because* it's garbage.

We stand now far above the shoulders of Giants, but if we do not take advantage of that we'll never see further than the tips of our noses. True, we are desensitized on the whole, but that should not be an excuse to wallow in deepening depravity, but rather an impetus to forge ahead. If we are so sophisticated now, why can't we prove it? What's become of subtlety? Where is the art with deeper meaning that *isn't* just mental masturbation? Can we not create pictures that truly *are* worth a thousand words?

I submit a rebuttal to all so-called art movements and manifestos of nihilism, a return to the rules with something of ourselves included, and for the sake of posterity name it the Oblique Movement. The word "oblique" is commonly defined as "slanting" but also as something "expressed indirectly, not straight to the point." An Oblique work of art would require subtlety and visual metaphors to communicate an idea; would be open to interpretation, but not frivolous interpretation; would rely on skill and technique with form and composition; and would actually communicate something. It must be complex enough to tell a story, but not so complex that it's difficult to view. And it must not fall into the same hands of those who quibble in semantics and claim their art is too lofty to be understood by any but the most enlightened (read: lemming-like). It must require actual thought on the part of the artist, and invoke thought on the part of the audience. What is art, all art, but a means of communication? Something must be expressed by the artist to the audience, or it is a waste of time and effort on both parts, and it must be expressed understandably rather than in the "Oh, you just don't get it" style of poseurs. If a work of art can be interpreted many different ways (ie. "It means different things to different people") then the artist has failed.

I am reminded of Paul's first epistle to the Corinthians, in which he says "Even things without life, whether flute or harp, when they make a sound, unless they make a distinction in the sounds, how will it be known what is piped or played? For if the trumpet makes an uncertain sound, who will prepare for battle? So likewise you, unless you utter by the tongue words easy to understand, how will it be known what is spoken?"

And this in turn calls to mind something a young woman once said of Jimi Hendrix, that "He was so far ahead of his time we may never understand him." What good is that, then?

One of these two speaks to me of Art, and the other just noise.

I posted this on several online forums, and was startled by some of the responses. One of the guys said that my manifesto was a travesty and I have the face of a mongoloid baby. I have trouble understanding how he reached that conclusion, but that was all he wrote. Or she, come to think of it. One thread is reprinted below:

Some guy who named himself DarkLord (Really? Come on) almost immediately posted:

Johnny Yen wrote:
What is art, all art, but a means of communication?
Are you talking only about art for public consumption?

On another note, I think using that definition of art might admit some things that the rest of your Manifesto would want to preclude ("White on White" for example).

Every art movement in history has been replaced with another art movement. Since the early 1950's what is left to do? Although the thread has focussed on visual art it's fascinating to cast an ear to music styles (movements?). Occasionally, if the attention is working well, it's amazing to hear recordings of music in which there occur short phrases that would much later (maybe generations) be the bread-and-butter of a particular style. It was potentially available then but there was no ear for it. The time had not arrived.

The third death "is that moment sometime in the future, when your name is spoken for the last time" - David Eagleman

To which I, perhaps too hastily, replied:

It's pretty clear. Where it says "What is art, all art, but a means of communication?" the words "all art" are there to mean "all art" and not "art for public consumption." Now, what do you mean by the "White on White" comment? I do not understand. If you are referring to Malevich's "White on White" and Suprematism, it is not admitted because it communicates nothing, aside from "If you don't play along with my little sham of a painting, then you are close-minded." It is a square. Period. A page on www.rollins.edu says this: "In his 1918 Suprematist Composition, White on White, a step forward from Yellow Quadrilateral on White painted a year earlier, Malevich attempted to eliminate all superfluous elements, including the color; since in 1918 he virtually gave up painting, perhaps these experiments convinced him that he had reached his goal and could not develop his Suprematist ideas any farther." I find that absurd. It is the praise by one poseur of another. "A step forward?" Hardly. This is what I mean by backsliding in the name of progress. What I propose is truly a step forward. Putting MORE effort into an artwork than the LESS that these others advocate would be a step forward. Not seeing how little effort you can get away with. And let me submit another maxim: "It's not art if the same work could have been produced by vandalism." That is to head the next detractor off at the pass. Such things are not art and it has been said often enough that they are expressions of one individual's blah blah blah about yadda yadda yadda. Making allowances for drivel does not make it more than drivel. Sticking feathers up your butt does not make you a chicken.

Then someone calling herself (or himself, come to think about it) LandLady, said:

When I was taking a philosophy of art class, we were introduced to the works of Zhu Yu that were presented at the *Fuck Off* art exhibition in Shanghai in 2000. I guess you can call his work Shock Art, with a moral twist. His works were, for the most part, rejected by the general public (similarly to *Piss Christ* and *The Holy Virgin*), but his ideas struck me as very original, as I've never seen art like that before. I am not an expert in art movements and I guess his art would mostly qualify as extension of conceptual art, rather than a new art movement by itself.

Anyways, there is this new movement that is trying to emerge; it's called Remodernism. It's not abstract in any sense; in fact, it seems to embrace realism and content.

It has emerged from a Stuckism movement, whose main goal is to re-establish spirituality in art. Remodernism seems to attempt to go back and replace Postmodernism as the next movement after Modernism. It claims that Postmodernists were too cynical and superficial and did not fulfill the potential of Modernism.

There is a sub-branch of Remodernism called Defastenism (though by looking at their manifesto, I am not sure how to tell the difference between Defastenists and Remodernists).

P.S.: There is also a counter-Remodernism movement which defends and upholds Post-Modernism.

I wasn't around to reply or even Google those terms for a few hours, and so these replies were made without me to participate. In that time, the Counter-Post-Remodernism people reared their ugly heads and had their fifteen minutes, then faded into obscurity, and probably spawned even shorter-lived splinter groups, like Abstract Interpretational Anti-formalists. But I liked what the next guy said:

becomingagodo wrote:
Every art movement in history has been replaced with another art movement. Since the early 1950's what is left to do...?

Maybe I am blind, but is their anything else left to find in art? Since the early 50s, the art world has had:
Pop Art
Op Art
Miminalist Art
Political Art
"Earth works"
Street Art and Graffiti
Feminist Art
Video Art
Digital Art
Globalization

Isn't that enough for you....?
Oh, and the idea that art movements smoothly succeed one

another is an invention and distortion of history.
Get thee to a class in contemporary art...

To which someone calling himself Machiveli replied:

We seem to have become stuck in a century long 'what is art?' debate - a purile game of transgression played between artists and critics. The only thing which hasn't changed is the economic and social system of the art world. That is the fetishisation and collection of objet d'art. Art has convinced itself that it has inherant worth -This must be stopped!

rules

1) Anyone who abides by these rules is part of the movement as an equal to any other.
2) All works must be submitted anonymously. all contibutors to the movement are prohibitted from claiming ownership of a work
3) No art objects produced by the movement may be owned.
4) No one may profit from the movement or work in any way either socialy or economically
5) Treating art objects in a way which implies that they are worthy of respect is banned.

I'm toying with the idea that:

6) Works must be accompanied by a short statement of purpose i.e. what was the intented effect of the piece.
7) Should the work not have it's intended effect on a significant number of of the intended audiance. The artist should be punished.

In short art needs to wear sackcloth and repent of it's sinful self indugance!

Wow. That's copied and pasted, by the way. I didn't make any of that up or write false misspellings. It was all *bona fide* bullshit. Followed by:

Explore the subcategories, redefine, transmutate, explore theory, explore the surface, deconstruct the image, exploit the potential. Use your imagination. There is infinite potential, and there are many more subcategories within these aforementioned categories to be explored.

Colleptic   --Slipping into the Illness

and

Don't lose faith in art. Just because a new movement isn't evident or the idea is not well known doesn't mean in some basement across the continent some guy isn't secretly changing art history. I agree with Colleptic whatever new movement begins they will find links and ties to older artists as means to describe the influence and origin. This question you ask of art could be asked of many thing. As long as people are alive and thinking there will continue to be new ideas or new ways to express old ideas. It's just a waiting game.

The only response I can think of is: Umm...
But then this woman shows up (Hallelujah!)

#1. Thanks for posting something interesting and challenging to discuss!

#2. If I can make a suggestion without seeming out of line, you might also take Paul's advice ("So likewise you, unless you utter by the tongue words easy to understand, how will it be known what is spoken?") and try for a bit more readability. Your manifesto is quite interesting but i'm afraid it's also quite hard to wade through, both in organization and vocabulary. I think it's unnecessarily so; I think quite a few people will start reading it, then bail out too early, leaving them with an incomplete understanding of what you're trying to say. For example:

"Anyone capable of free thought who calls this exactly what it is, well, merely doesn't get it."

Ummm, I don't get what you mean by that sentence. sorry.

#3. On to the point you're trying to make - which is I think quite interesting, and that's why I'd like to try to paraphrase what you just said a little more simply. To put it very succinctly, it sounds like you are objecting to several aspects of the modern art establishment, especially the tendency to rebel against the influence of previous generations. I have to say that that's also what you're doing - rebelling against the previous generation of modern artists, such as Warhol. Right? Now, I don't think you're doing it just to be controversial, I think you're doing it because you see value to the methods Warhol and his contemporaries threw out disrespectfully. But my point is that the history of human art is a pattern of responding to the status quo and rebelling against it in different ways. I don't think that's a bad historical pattern - it gets people thinking and disagreeing, exactly like your manifesto!

So instead of objecting to that pattern of rebellion and reconsideration, which I think is pretty much inevitable, I think your best point is that the last generation didn't value the good parts of the techniques and traditions of the old masters. (I hope you have the ARC's website (art renewal center) bookmarked, sounds like you'd love it). I'm a big fan of the old masters, but I'm kind of glad that modern art is out there too.

Some modern masters can be both surprisingly oblique and subtle, and surprisingly beautiful. The mobile sculptures of Calder, for example, blow my mind almost as much as a Klimt painting does.

And I have to disagree that all the good techniques and aspects of art have already been discovered, as you seem to imply. I think there is always something left to discover in the process of doing art.

Upon seeing the cave paintings of Altamira, Picasso said that everything since was nothing but decadence. That's a more extreme view of the perspective you espouse: that art progressed up to a certain point, then became corrupt and stopped advancing. Picasso just puts his zenith of art a few thousand years before you do. But I don't agree with either of you - I think there is always something new left to discover in art. Why? Because art is a dialogue with society, and society can't go backward in time, it just keeps building on itself.

Quote:

What is art, all art, but a means of communication? Something must be expressed by the artist to the audience, or it is a waste of time and effort on both parts

I completely agree with this statement.

jess
Frogbells

And then this guy:

There are two arguments within me. Both for and against.

While some contemporary and recent art is nothing but the drabs of charlatans supported by incapable critics; That is not to say that contemporary art trends are without merit but equally well it doesn't make them the cats pyjamas.

I respectfully disagree that everything has been tried and discarded, art is never stagnant, we have materials today that the old masters never had access to. Many people said photography would be the death of painting yet, it has adapted and changed to cope with our modern world for what would a world without art be but a stale environment, if we throw out our artistic sides then we impoverish ourselves but we should not be so quick to declare what has and hasn't merit in our eyes.

In it's own way your statements can be seen as a blindness as well, you're groping for the past in a starry eye assumption that things were better back then and those who called themselves artists were always skilled.

All form of art, painting, poetry, literature do well to remember the lessons of the past but there is a danger in becoming too hide bound that stranglation of the future may develope. Who wishes to live in a world where every artwork is "perfect" where every written story or poem is grammatically perfect but all are as dry and as dead as 1,000 year old bones because technique was concentrated on to the exclusion of the ellusive quality that turns words into art.

Art isn't just about technique, all the technique in the world won't save an artist who has nothing to say. It's a delicate balance of emotion and skill. An endless conversation with the viewer or reader in which everyone finds something new just for them.

You might dress it up in fancy words but ultimately I think

you are as blind as those who advocate everything is art. One does not throw the baby out with the bathwater, one pans for gold whether it is of the esoterical variety or more indiscriminate. Not everything made by an artist is art but equally well not everything outside of a small section is necessarily trash.

I find it hard to take many of these critiques seriously because the writers didn't even try to compose coherent sentences or spell the words that comprised them correctly. It's not like every computer does not come with Spell Check. You could at least *try*.

I blew off the other guys and wrote:

You're right, Jess. Everything you said in your response was spot on. The problem with the "merely doesn't get it" line was that the message board wouldn't accept it as it was originally written. I tried posting it almost 20 times, and had to keep going in and editing little things out, like quotation marks. There were supposed to be quotes around that part, showing it was the attitude of poseurs saying "Oh, you just don't get it" whenever someone criticized them. I had to change it so many times I got fed up and took every quotation mark out, along with every instance of my personal writing style. The effect of it was changed dramatically, I think, and not for the better. But hey, at least the damned thing is posted finally, and generating responses. Thank you for replying.

SI wrote:

While some contemporary and recent art is nothing but the drabs of charlatans supported by incapable critics; That is not to say that contemporary art trends are without merit but equally well it doesn't make them the cats pyjamas.

Agreed.

Both extremes I find distasteful - this manifesto is way too extreme, and way too dismissive of what is recognised the world over as great art of the 20th Century - sure if you *want* to write a harsh critique of Warhol or Basquiat that's eminently possible. To dismiss it as "garbage" is just an disgruntled artschool flight of fancy. Jimi Hendrix - noise? Get real. In saying that you go so far out on a limb as to not be credible.

Equally on the other side, the total reduction of earlier centuries art to museum pieces and the reduction of craft to a quaint sideline irks me too. The reaction against academic craft I can understand and buy into to a degree - but not to the extent of anti-craft, often taught in artschools, which all seems rather reactionary to my mind.

Luckily both types of art can happiy exist, and do so together in museums and galleries the world over. Its a wonderful thing to be blessed with an open enough mind to enjoy both contemporary art and old masters. To have the patience to look for reasons why the entire art establishment disagrees with your personal view is a handy skill too.

Certainly the rules of the Salon, when that ruled the artworld with hard classical rules, produced some ponderous, overblown crap. As do some current trends. But amongst all that there's great stuff to look at and learn from in every era of art production.

To dismiss a whole century of art history is frankly a rather silly, reactionary propositon. Just as any attempt to discredit older centuries art is much the same.

Then some nitwit calling herself "Stained Aprons" wrote that she makes installations and felt that painting is dead. I asked her what she meant by that and she said "Just that.

Painting is dead."
"In what way?" I asked.
"Well I think it's pretty obvious what dead means. Lol"
"But painting seems to be alive and well to me. How does it seem dead to you?"
"Come on just look around you. Painting is dead."
Then I realized she was an idiot and I let it drop.
The other guy, I had to respond to his Hendrix comment. Not be credible, my ass.

Woe is me, I've been misunderstood.

The comment about Jimi Hendrix was not so much about his noise, but a response to what the young lady said. That he was so far ahead of his time we may never understand him. It's that way of thinking that I object to. And I wasn't saying there are no contemporary masters. Indeed, there are many, and I have no illusions about ever ranking among them. But calling the ones I've named (and quite a few others) Masters, frankly, pisses me off. The NEA putting a jar of urine with a crucifix in it in a museum and giving that swine a grant pisses me off, and this is my rebuttal to him and all others like him. Calling Hendrix noise, though, even if I did mean that, is not going out on a limb. His mangling of the Star-Spangled Banner was noise, pure and simple.

Somebody else said:

I always found the emotional art school discussions tedious. I had really hoped that folks had grown up a bit in art school and that age old argument was over.
I find the whole reverse snobbery thing a bore. Illustration isn't art...this isn't art..that isn't art...but my throwing of buckets of paint at a canvas and hoping I get something pleasing is ART. My take was at the time..."Well...what the hell am I doing here paying for this course? I don't need lessons in paint throwing."

Not that there isn't some good art that is created this way or very pleasing to the eye...and therefore it is art. So don't think I am saying it isn't art, but a whole lot of it ISN'T

I also suspect that those who do create pleasing art in a revolutionary manor do have the basics in art down. They would have to. You learn anatomy before you stylise...you learn color theory before you throw paint.

Perhaps one of the reasons that many of my former classmates who argued that illustration wasn't art are now accountants and haven't even looked at paint or art in the past 30 years and I have been an artist for 30 years...selling my work and making a profession out of it.

Charlatans could sell work by Game. Seriously I saw it a lot in my early days of trying to break into gallery art.

There on the wall of a posh gallery would be the most godawful mess of paint. With a price tag on it that was outrageous. I would stand there going "HUH?" Then here comes the artist accompanied by the gallery owner...in a damn costume...I would laugh...shake my head....wonder why he hadn't cut off his ear so he could really look the part.

The artist was the star...not the work. The work was mediocre at best. But the public wanted it because he had a publicist who sold him... They would make up stories about the artist....just complete crazy off the wall nuts stuff...and the public thought that was art. Nearly every time the artist had an accent that just sounded made up to me, Too MUCH.

One bearded jerk who couldn't paint his way out of a paper sack said he was a Russian Baron...I was taking a bet he was really Joe from Brooklyn. I use to think they had a casting call for artist types that the public would buy and then they gave the person that fit the type some quick lessons in paint throwing. I thought the public was just plain Stupid...I mean really.

I grew up around art and illustrators...Good ones...famous folk. And I knew they were just normal everyday people. Who were these weirdos that they were selling as artists to the public? I actually thought about ways I could sell myself. Thinking hmmm...what can I do? I am not 100 like Grandma Moses...and I am not a guy. I can't grow a beard....although a bearded lady might have made it I think. I mean really they wanted a Gimmick to sell you...The art wasn't really that important. I didn't have a one...I am female and wholesome...I was screwed. Every single gallery turned me down...I as a person wasn't sellable....the art was great...they liked it...but they didn't see anything that could make me famous....I was told over and over. "You don't look like an artist....you look like a fluffy little blond cheerleader."

Actually they weren't that nice...they really stuck their noses up...like I was something smelly and slammed doors...a few times when the art was brought back to them without me...I heard. "I love it." and then they walked over to meet me and the face fell...Can't use her.

I was told to go paint flowers on lampshades and birthday cards or fashion art...or arts and crafts. Females did over the couch work...and in illustration...it was ok if you did Fairies and cutesy and pin up girls.

For the most part that isn't true anymore. Art is being loved for art and the public is learning what is art and not looking to buy art because the artist had an affair with Marilyn Monroe, or Rock Hudson, or he cut off his ear, went on a hunger strike and lives in a loft with a monkey. So I think personally. If art schools are still doing that same old bull...let them go broke. That will eventually teach them.

Kay

"Fred, you are next." says the art instructor. "Show us your project."

"Here is a picture that I did last night." Fred holds up a big canvas with 3 big splashes of pink, blue, and yellow paint. "This represents three parts of our psyche...id, ego, and the superego."

The art instructor applauds. "Very good, Fred! Now, let's see yours, Julie."

Julie stands up and holds up a block of wood. "I found this block of wood in a ditch. To me, it represents our struggle against nature. The wood itself is nature, while the sawed off ends emphasize our fight to conquer and control nature."

"This is just perfect!" the art instructor wrings his hands in delight. "I'll have to tell our school to include this in our gallery. Susan, you are next!"

Susan stands up and holds up a beautiful painting of a dog lying in a bed of flowers. "This is a portrait of my dog, Fido. It took me over a week just to capture her fur. I worked hard to make the picture as realistic as possible."

"What's this? A picture of a dog!" the art instructor didn't seem pleased. "This is art class. We don't do portraits of animals or people here."

"But..."

"You need to loosen up! You need to expand your mind! You need to reach into your soul and give us something meaningful!" The art instructor cried.

"But..."

"Sit down, Susan. We will talk after class. Ok, Bob, what do you have here?"

"I have a jar." Bob holds up a jar.

"Yes, what's in there? Looks like it is full of lint and hair."

"Well, that's my bellybutton lint collection. Every evening for a decade, I pulled lint out of my bellybutton and placed it into the jar. It represents my frustration at the industrialized society."

"Wonderful! Wonderful!" the art instructor cheers. "I'm

going to have to recommend you for a scholarship!"

-Mike

That is it Mike. Exactly. Love it.
Belly Button Fuzz...yupper.

At the risk of really, really pissing people off, Mike, I don't think that art teacher is way out of line. You mean him as a joke, but I actually agree with him.

The fact is, a painting of a dog is a painting of a dog. You don't specify what exactly the painting looks like - but if the painting is nothing but a meticulous reproduction of a photo of the dog down to the last hair, it is am impressive example of skill and technique, but it has no creativity or thought behind it. How is it in any way better than the original photograph?

The other works in your example sound as if they're done by people with no practical abilities, but who can "talk the talk." A cynical person might read your story and say none of the other students actually believes what they're saying, they are just "selling" their art to the instructor. If that's true, it's depressing. But suppose they actually are trying to say something about Freud's interpretation of the human mind, or our struggle against nature? At least there's something to think about there.

In your example, it's all extremes. But in life, most artworks are balanced between these two extremes. I personally think an exciting artwork is about showing you something you haven't seen before, making you see something in a new way. It's provocative. A painting of a dog CAN be provocative.

The dog can represent many things. The artist can choose to emphasize aspects of the dog's personality or what the dog represents to its owner. A great portrait artist ALWAYS gets a result that a plain old camera could not have captured. And in education, illustrators are important because they can manipulate perspective, color, and detail to acheive something clearer and more informative than a photo. (And there's photo-artists who manipulate their camera to get very impressive results - a camera is a tool like any other, but the key here, is an artist isn't just point-and-clicking, they're thinking and composing).

If the assignment in your little parable was to create an artwork that embodies one of the struggles of society in the 20th century, then the other students did a better job at the assignment than the dog painter. You could definitely mark them down for being lazy and not having/showing any technical skills, but at least they had a context for their art object. (Obviously if the assignment was to demonstrate your skill in executing precise representational detail, the dog painter won).

So how is it possible that the quality of the "art" varies with the assignment of the art teacher? Shouldn't "art" stand on its merits no matter what that silly old teacher asked the students to do? Well, yes - ideally the message of the art would come across to the viewer without a little caption or museum docent there to explain it to you. Truly great art doesn't need someone to stand there all the time and say "this is about industrialism." But then, truly great art is rarely viewed in a vacuum. If we know something about its context, that enriches the experience of the art. Guernica is impressive, but once you know what it's about, it's REALLY impressive.

Like it or not, the meaning and concept behind the art, what it gets you to think about, is a big part of art (people seem to forget that the Old Masters put massive quantities of symbolism, Biblical references, and political meaning in their works).

"Art" with no meaning or implications at all is a Thomas Kinkade painting.

Or, a painting of a dog. Have I rocked the boat enough?

> Frogbells

>> That's a difficult one. I agree about the dog not really embracing the spirit of what artwork should be (assuming it's just a base portrait) but in this extreme example I still think it triumphs over the other contributions. No matter how inane an artistic effort may be, if any degree of talent above the norm was employed to create it then it will *always* triumph over the pickled dicks and breeze-block heaps of this world. Talent and meaning should combine in an artwork to provoke strong emotion.
>>
>> When it comes to peer pressure about what is art and what isn't, and what is *good* art and what isn't, I have my own ideas and I'm sticking to them. I have no problems in proclaiming entire award-winning galleries of "art" as garbage, in a similar way that I can say Van Gogh was an appalling painter who didn't even deserve to sell the single painting he managed to offload before he died. And as long as I know what I like, I'm happy
>> --Baron Impossible

> I stand with Jessica on this one.

Really, art has ALWAYS been about the sharing of emotions, thoughts and physical sensations between the artist and the viewer. Beauty, or the feeling of beauty is one of these, but by no means exclusive. And as a matter of fact, many classical masters were not after "beauty", or in a very broad sense then (like the "beauty" of a shared emotion, even if it is intrinsically unpleasant). Think of the tortured body of Chirst in Grünewald's Insenheim Tryptic (that would make Mel Gibson's movie look like a children TV show), or some the most desparate self-portraits of Rembrandt, and you will see what I mean. You do not need to go into modern art to see pieces which convey a feeling of uneasiness, sometimes of ugliness or even pain.

Now, what is wrong in contemporary art (assuming there is something wrong) ? My opinion is that it has all become contextual. Jessica mentioned that point already: no work of art is devoid of a context, and most often that context (and one's knowledge of it) is a key to "enter" the work. But the german poet Reiner Maria Rilke said it very well, when he spoke of "the loneliness of the work of art". What he meant by that was that one should always confront a work of art trying to strip oneself from his/her culture, trying to strip the work from its context too, and face it as if it was alone and without a time reference. To me (and to Rilke), a real work of art should stand that test: even without contextual reference, it should still be able to create an emotional, sensorial or intellectual response in the viewer's mind. And then only should context intervene, and still reinforce and deepen that communication.

The problem with contemporary art is that it is often completely contextual (and in a voluntary manner). Art students are TAUGHT to make statements, they are TAUGHT to forget about the "loneliness of the work of art". And there lies the divorce with the general public, in my opinion. Most of these artists are not quacks. They are sincere in the path they follow. Only, they work according to a system of values where that "path" they follow has become THE valuable thing in their art, while the individual pieces are seen as just landmarks along that path. Hence, without a notice, without a contextual information on the "path", without the knowledge of all those previous landmarks, the pieces become void, meaningless. And from there comes the general public misunderstanding, like "this piece of wood is not art, it is crap". A piece of wood in a gallery is crap, by itself. But if you grant value to the context, and know why this piece of wood is there, then it can become art. But to me, it is an art that has lost all universalism, an art that has become ALL culture. And for that I feel it has failed.

Now, do not think that all contemporary art is like that. There are many trends that still aim at keeping that intrinsic non-cultural connection with the public (even if, again, ANY work of art is supported by culture, whatever the time it was made). One art movement I really love and which has retained its universality (in my views) is Land Art (which you could, in a way, describe like "landscape-shaping"). If you have ever seen works by Nils Udo, or the incredible pieces of Andy Goldsworthy on the passing of time, then you will again see what I mean. In these works, you will find "organic" qualities which are intrinsic, and so much linked to our senses (in their basic biological definition) that any human being can relate to them. And at the same time, they are not just "pleasant" settings: they convey very strong and thought-provoking things about human condition, and its often futile and temporary nature (especially for Goldsworthy).

So really, I understand and often share the frustration of the public with contemporary art. But too many people reject it for the wrong reasons, for lack of understanding what it really is. Alright, now you can call me pretentious, because I claim to reject most of contemporary art for the RIGHT reasons !

Best,
Pierre Carles

Ok, I think I get it now - Johnny you're defining craft as art. EG: If something is well made then it is art.

A great many artists (amongst various other types of people) over many decades don't agree with you.

About Hendrix - here's the thing - his starspangled banner was not meant to be a melodious beautiful rendition of a proud national anthem. It was meant to be a destruction of that national anthem to reflect his feelings about what was going on in Vietnam at the time. The guy can play fantastic guitar, and yet he's on stage demolishing this well known tune... that says something, and was an incredible performance.

If you chose to judge it solely in one way, without considering its context or intention then yup, you will hate it, and see it as noise. Just that same as if you judge a Pollock by Classical standards.

The way something is made can also be its content. Personally I find that an intriguing and engaging idea. Form can actually be function. That's very much a core principle of Modernism. Now, I can also rip that idea apart if you like, having read enough post modernism to do a fairly comprehensive job on it...Once again in this kind of debate Pierre has covered a lot of what I would like to get across much better than I can. And Jess too.

I think what chafes me most personally is this idea that all these great artists of the 20th Century were quacks or snake oil salesmen. Read into the lives of these people a little, and you'll see that's absolutely not true. Some of them died for their work. Some of them fled halfway round the world to escape Hitler or Stalin's repression of modern art. This was serious stuff to them, and far more than just some con.

In addition to my commerical work I paint abstracts myself. I'm not conning anyone. I really love the idea of painting without trying to render observed forms, or follow precise rules of rendering laid down by some guy in the 1800s. Those rules are not the be all and end all of art for me, and I'm really happy with that.

DarkLord

Ahh, the Dark Lord again. Did anyone ever say craft was art in this debate? Did anyone ever say that craft was defined as something well-made? I did not bother to reply. All I could think of was that he really loved the idea of painting without trying to render observed forms, and he's really happy with it.

I wanted to tell him that sometimes I masturbate too, but I don't have the effrontery to call it sex. I didn't, though, because that would have made me no better than the guy with the mongoloid comment.

I left all the discussions on all of the forums because they were all variations of the same. Except one anonymous contributor who said that he cried tears of joy that I'd written my manifesto and restored his faith in contemporary artists.

I told Elizabeth about it later and she was excited. She sat down and read everything on her laptop—and I mean everything—from each forum, laughing here and there, nodding a bit and groaning from time to time. I did enjoy the attention she paid to it, even if it did feel a little awkward to be sitting there doing nothing the whole time.

Finally, she told me I had to come to Antigua because it was an artist's paradise. There were art galleries on practically every corner, and with all my ideas and talent, I'd become a god in no time.

And for the first time, I actually considered it.

VI

On one of those days, I took her around to meet some of my friends—Rene, the guy I mentioned earlier with whom one did not fuck, his girlfriend Nicole, this chick Demi (who was wicked smaht and funny as hell) that I've known since middle school, and Nick and Tate and them. They'd gotten it into their heads that the way to get to know someone new was to sit down together in a dark room and watch a movie, and thereby interact in no way whatsoever. At least, that's the way it seemed because instead of sitting around drinking and talking, they said they'd gotten the director's cut of Blade Runner and we all had to watch it.

"What's Blade Runner?" Elizabeth asked.

They all looked at me, saying nothing. As if they were judging her by that, which made me feel uncomfortable. Doubly so, because I didn't know either.

"A futuristic Harrison Ford movie about hunting renegade androids," Demi said.

"Oh. Sci-fi, then."

"*Great* movie!" Nicole chimed in. "*Ex*cellent film. And it was Rob Zombie's inspiration for *More Human than Human!*"

"Who's Rob Zombie?" Elizabeth asked.

They all looked at me again.

Their judgment of her wasn't lost on Elizabeth, but she let it slide for the moment, and I knew it was going to be held against me later.

Time out.

There's something I ought to mention. I have a habit of making all kinds of different friends. I have preppie friends who own boats and have other people mow their lawns, I have illegal immigrant friends, I have punk rocker friends, and then I have my smart friends. I love them all equally, and only once have I tried to invite them all to a party to get to know one another. It was a frickin' disaster. They all sneered at each other and criticized me for hanging out with any other type of person than themselves. Well, except for Rene and Nicole and Demi, and Nick and Tate and them.

They've always been the most easygoing and the most welcoming of others. Trust me, everybody likes them within minutes of meeting them. They are the smart friends, by the way. So this was weird.

Time in.

We got our drinks and settled in on the couches, and the movie came on. If you haven't seen it, it's a movie about androids who work as servants and slaves and have a life span of four years, and some of them escape a planet where they were supposed to be mining something or other. They return home to Earth, and are trying to find the scientist who created them and get him to give them more time before they shut down.

It's Harrison Ford's job to find and kill them all, and in the process he meets this female android who doesn't know that she isn't real, breaks her heart by telling her she's an android instead of a real human being, and then falls in love with her (as ya do). Then the two androids played by Rutger Hauer and Darryl Hannah arrive and fuck everything up.

There are a couple of lines in the movie that really struck me. They're worth repeating, I think.

One of them was near the end, Rutger Hauer telling Harrison Ford about his mortality, saying that he'd seen attack ships on fire off the shoulder of Orion, and c-beams glittering near Tannhauser Gate, and how all that would soon be lost, like tears in rain. And the other was Edward James Olmos being a detective and referring to Ford's little love affair with an android, saying it was too bad she wouldn't live, but then again, who *does?*

That really hit me. Who do we think we are with all these plans that we make and the lives we put everything into, these lives that are just going to end one day? Everything we do that we call important, it's all so trivial when you think of how we're scurrying around on the surface of a rock, a rock that's just spinning around and around a ball of flaming gas. We've no idea how we got here, no idea where we're going or what we're going to do when we get there, or why we're even here in the first place, and so many of us are killing one another because of disagreements over the answers to these questions. And where does it all get us? Nowhere. All of us die, and in dying, are eventually forgotten.

So why the hell should I care if some of these ants that I share space on a rock with don't like the ant I'm sleeping with because she's never heard of some other ant's music, some other ant that none of us ever met, or ever will?

The rest of the night I felt very detached.

Fast-forward a couple days, after the touchiness about how my friends reacted to Elizabeth passed and I was talking with Demi at the Starbucks where Nicole worked. I had come out and asked the two of them what they thought of my new girlfriend. Nicole excused herself to go serve somebody a double-half soy latte macchiato with a twist of lemon, and Demi beat around the bush for a minute until Nicole got back, lit a cigarette, and said "She's a bitch."

Demi nodded. "She looked at us with her head tilted back and a little bit sideways, and she blinked a lot. Also, she sat with her arms crossed, and when she shook all our hands, she was palm down."

"You're kidding."

"Nope. She did all that."

"So what?"

Demi sighed. "You're not a woman, Johnny."

"I am well aware of that."

"Women notice that kind of thing," Nicole said.

"Well, maybe she was doing all that, whatever it means, because you were all acting cold to her when she didn't know who Rob Zombie was. You ever think of that? You did all get cold right away."

"Johnny. Do you really think we're that shallow?"

"I didn't, not until that happened."

"You weren't paying attention at all, were you?"

"To what?"

Another sigh from Demi. "We'd never be cold to someone because they didn't know something about music or an eighties film. There's a lot out there that *we* don't know about to go around judging others based on what they don't know."

"Speak for yourself," Nicole said.

"Then why'd you get all quiet when she asked 'What's Blade Runner?' and 'Who's Rob Zombie?'"

"It wasn't what she said. It was how she said it."

"...What?"

"She asked what Blade Runner was with her eyebrows raised and her eyes half-closed, with a sneer, like "What's *Blade Runner?*" like "And that was in the Cannes Film Festival *when?* Like what*everrr*.""

"She doesn't talk like that."

"You just think she doesn't because you're fucking her. Trust me, she does."

"Yeah," Demi said. "She sized us all up and dismissed us like we're nobodies. People she just has to tolerate for a few hours before she can get back to report our habits and behavior to her alien leaders."

"Good one, Deems," Nicole said, and high-fived her. I drank my coffee and changed the subject.

Now, rewind to the day after watching the movie, and we won't bother about the bitchiness because it'll be as tedious in the retelling as it was to experience.

I started painting the portrait. Now, like I said before, Elizabeth has alabaster skin because she, like her grandmother, stays out of the sun to protect it and keep it flawless. The thing is, it's attractive in real life but visually un-stimulating in a photo or, I'm finding out, a painting.

If you want to paint a nice portrait, the subject has to have (if she is white, anyway) at least twelve different colors to her skin, which makes calling her "white" rather silly, in my opinion. You look at any white person and they're going to have a lot of pink, a little purple, a bit of terra cotta, and even some green in their "beige" faces. This is why it never looks right when someone just uses the paint out of the tube called 'Flesh Tone.' So, I painted her as if she were a normal, multihued young woman.

The dress was tough. Because of all the gaps in the material, making her more nude than clothed, I had to paint her nude first and then crosshatch the dress over her, and doing the shadows for every individual gap was a pain in the ass.

For the background, I just did one of those formless dances of color and light that I see in so many other paintings, choosing warm and neutral colors that comforted the eye and induced feelings of peace.

And Agnes hated it.

"What's all this color?" she asked. "Liz is *white*."

"Well, yes, she is, but most portraits are painted to a romantic ideal instead of portraying the exact idiosyncrasies of the subject, because of that subject having their vanity—"

"Don't try that double-talk with me, sonny boy."

"It's not double-talk, Mrs. Morrow. I've done this for years. I used to paint people exactly how they looked, and they got offended that I made them have dark circles under their eyes, or freckles, or their muscles weren't big enough or their gut was *too* big. I have learned to overlook all those perfectly-normal attributes, and wasn't about to make the woman I love go through the self-consciousness of pointing out what she hates about herself, like many others have."

Agnes watched me shrewdly for a moment, with those huge owl's eyes behind Coke-bottle glasses.

Finally, "The woman you love, eh?"

That caught me off guard, but yeah, I'd said it. Hmm, Freud, c'mere and explain this one to me.

I started looking at Elizabeth differently from then on, more critically. I remembered something that Demi had once said to me over drinks at Respect's.

"You're an infatuation junkie," she said.

"A what?"

"An infatuation junkie."

"I'd heard you the first time. What's that mean?"

"Think about it. Every relationship you have lasts about two-and-a-half, three months, right? Just enough time for the newness to wear off, and for that tingly lovey-dovey butterflies-in-your-tummy feeling to go away. About the time the sex stops being every time you see her, and the constant fondling and the way you hold hands, fingertips tracing patterns across each other's knuckles, and generally being revolting to everyone else in the vicinity. As soon as you start taking each other for granted, you lose interest, don't you? Because that feeling is gone. And so what do you do then? Break it off with her and go find someone else, and start the cycle again. That's what I'm talking about. You're always jonesing for that teenage love-sickness, and that's where you get your romance fix. You're an infatuation junkie."

"...I'm confused. Do you want me to admit to it and turn over a new leaf, or what?"

"Just tell me I'm right."

"Never."

But now that I'm thinking about it...I've noticed these periods in my life when I've started to feel that infatuation begin to fade, and forced myself to plunge my mind back into it, so to speak. Felt myself growing distant from someone and made a conscious effort to "love" her again, and that's when it started to crumble and fall apart. I became clingy with someone I was about to break up with, and then after she went and dumped me instead, I was so heartbroken I could have died. Which was odd. It happened at least five times in my life.

And now, as I was analyzing Elizabeth, taking note of aspects I wasn't exactly crazy about, I felt the threads of love I had started to feel for her starting to unravel, and even as I consciously noticed it, and was fully aware of what a bad idea it was, and could even make the most lucid commentary about it then as I am now, I still purposely drove myself toward the brink and over it, falling ever so stupidly for her.

I thought for a moment about a Guatemalan girl named Rocío, and Mayté, the beautiful Colombian in high school, and Amy the petite blonde from college, and Christina in Ireland. With all of these girls I overcompensated when my love began to slip away, and I forced myself to become whipped.

Something I'm ashamed of but can never undo, and pray that I never repeat, but know I am going to. Right now, by the looks of it.

"And you need to see Antigua in the spring," she was saying. "The jacarandas litter the cobblestones with their lavender flowers, and the purple and fuchsia bougainvilleas dapple the brilliantly-colored but crumbling walls of the buildings with—"

"Are you trying to make me come with you?"

"—what?"

"All of this bright imagery you're conjuring up, and the way you keep talking about 'when I come' as if it's carved in stone that I'm going to. All this talk about the future. What are you getting at?" I asked.

She faltered a little.

"Well, I guess I—no, I'm just telling you about where I live, really. It's just talk. I mean, why—"

"It's just that it feels like you're trying to talk me into moving down there. Just packing up and moving to some place I've never seen and leave all this behind and let the chips fall where they may."

"Are you saying you don't want to? Because if you are, then what are we doing—"

"I'm not saying that at all."

"Then what *are* you saying?"

Pause for a moment at the brink, looking over.

"I'm saying I want to."

"Want to what?"

Last chance. Think about it. Don't be too hasty.

"Move to Antigua."

Aaaaaand jump.

She looked at me for a moment, nothing about her body language registering shock or surprise, but everything in the air between us buzzing with both. Her smile was frozen, her eyes holding a lot of noise behind them, just beneath the surface. I could tell it was there, but couldn't tell what it was, exactly. Whatever it was, it certainly wasn't Yay! or That's great!

And if you've ever been on stage, you'd know what I mean when I say something took a performer's eon—that instant between finishing a set or delivering a punchline and the audience's applause—that split second that seems to last forever—the moment that all your insecurities come howling up out of that deep, dark oubliette of your subconscious, threatening to tear you apart. That moment that the die is cast and everything hinges upon how it will land.

I felt her look me over again, readjusting her opinion of me. In spite of being stoned—or probably because of it—I could tell that she had only pictured me in Antigua in this hypothetical universe, and now she was moving tectonic plates inside her head and transplanting me from here, from whatever this little temporary Palm Beach vacation-escape-affair was, and putting me there in the world where she lived every day, interacting with her friends and colleagues and accomplices, and seeing how I would affect her little world there and whatever she had in the mix. What would other people think of me, what would they think of her when I was standing next to her? All this *what? what? what?* was cycling through her.

And then, after that performer's eon, she caught up with it and hit the ground running, so to speak.

"Well, first, I'll have to get you a Spanish tutor. I know just the guy. He's a*dor*able. His name is Paolo and it'll cost you the equivalent of eighty dollars a week, which is really quite a deal. You *will* need to learn Spanish, yunno, despite the large expatriate population. It's funny, actually, in Guatemala a lot of place-names have this suffix 'tenango' which, in the local vernacular means 'place of' such as Jocotenango or Quetzaltenango, the places of the jocote or the quetzal, respectively. A jocote is a local fruit and the quetzal is the national bird, which, incidentally is the unit of currency there, the quetzal. Well, since Jocotenango is the place of jocotes and Chichicastenango is the place of chichicastes, and Quetzaltenango is the place of quetzales, we joke that Antigua is Gringotenango."

Pause for laughter. Her turn to suffer that performer's eon, multiplied by the fact that we're stoned and every instant is a hundred years long. That and I don't know what "Gringo" means. She sighed.

"Gringo is the derogatory term for anyone north of the border, usually North Americans, but also to mean Europeans nowadays," she explained. "*Used* to be derogatory anyway, like 'nigger' used to be but now's used freely by the people of that persuasion, and in the same sense 'gringo' is used in an affectionately derogatory way by expatriates living there. It reminds me of a joke. Okay, Guatemaltecos are known by their affectionately derogatory term 'chapin' and Costa Ricans are called 'ticos' and Salvadoreñans are 'guanacos' and Puerto Ricans are 'boriquas', but what do you call Mexicans?"

She waited for me to think, and shrug, and she grinned her cutely crooked teeth at me.

"Pendejos."

Blank expression from me. Her smile dropped.

"You'll really need to learn Spanish," she said.

Okay, so skip ahead, skip ahead, skip ahead.

I finished the painting. Elizabeth was there for the last few touches, helping me do it the way Agnes would've wanted me to do it, instead of doing it right. At least, 'right' in my opinion. She insisted that I put half of her nipple showing in one of the gaps of her net-like dress, but I knew that's what everybody would look for, and be disappointed in a way if it wasn't there and disappointed in me if it was.

I decided to err on the side of discretion, which seemed to disappoint her, but so what? I'd read that a woman will be a lot more disappointed in a man if he chooses to follow her idea that he disagrees with, instead of standing by his own convictions, even if he's wrong. That nonsense about a woman always wanting her own way? Bullshit. She wants a man who'll be a man.

So, I left the nipple out, and when she called me a pussy I told her you are what you eat, and grabbed her forcefully. She squealed with laughter while I hoisted her up and draped her over my shoulder to carry her caveman-like out of my studio and into the bedroom. I dumped her onto the bed aaaand skip ahead skip ahead.

But this part I have to tell you even though you probably don't want to read it. It's pivotal so I'll say it.

She got all serious at one point and looked up at me with her breath all heavy, strands of hair across her face and held there by the sheen of sweat.

She said, "I've wanted you to come inside of me, from the moment I met you."

Whoa, record skipping.

"It's safe," she said. "I won't get pregnant. But you have to promise me you've been tested."

"I have," I said.

"I mean recently."

"I have. I'm always careful about that."

"Good." She got on her hands and knees and presented herself to me. "I want you to nail the fuck outta me and come inside this time."

I had never done that before. I'd always been safe and pulled out, even if I was wearing a condom. This was a giant step for me, taking this risk with someone. It represented a willingness to commit, at least to me, because it showed a willingness to take a risk and get accidentally pregnant and stay together and make a future instead of just have sex for a while and quit. It represented her fuller acceptance of me as her man if that makes any sense. So I did it.

It was incredible. Really. This probably has happened to you a thousand times, but not to me, so it was like losing my virginity all over again. The first time, a girl thinking enough of me to let me in like that, which meant more to me than to her, and then that moment getting lost among so many others like tears in rain, and now this. I felt like everything I had ever done had been leading up to this moment.

Which was funny I guess, because moments like these are the most perfectly natural thing in the world. I mean, there wouldn't be anyone on the planet without it happening all the time, everywhere. But I'm romantic, so let me have my romantic little moment.

We lay together for a long time, holding each other and listening to the rain outside my window. I hadn't even noticed when it had started.

Remember what I'd said earlier? About how it all seems trivial, when you think of how we're scurrying around on the surface of a rock, that's spinning around and around a ball of flaming gas? When it comes right down to it, there's only one thing on this rock that matters: This. A strong male like me coming inside of intelligent females like her. Or maybe it was the weed that magnified the significance of such a simple and common thing to my overly-imaginative mind. Whatever.

I was in love.

We made plans the next day, and ordered plane tickets online. Since I, as a tourist, would only be able to stay for three months and then apply for a visa extension, we made my return date three months and two weeks from that date. She said two weeks ought to be enough time to get everything situated on both our ends. I could get all my paintings out of the art galleries they were in—hey, I really did have some in a few of the venues around town, just not as many as I wanted—and have them shipped to Antigua. Rene would help me out with that because he worked at FedEx and had an employee discount that made the rates go waaaaay down.

The biggest problem would be telling everybody I was going. That part made me uneasy, but I bit the bullet and told my family first. They either congratulated me or cautioned me, but all wished me well, and it made me happy to know I'd be missed. Well, except for my Mom. I feel I have to mention my mother's reaction.

The first thing out of her mouth was "I'll be all right."

That bothered me. She didn't say "Really?" or "Are you sure?" or "That's great!"

I told her and she paused a second, then said "I'll be all right," as if she were trying to convince herself of it, and that made me realize just how much she depended on me. To turn around when I got home at three in the morning and walk out the door to the 7-11 nearby, the one she could have gone to at any time, and get her more cigarettes. To get all the groceries. To do pretty much everything so she can stay in her seat at the table and do her crossword puzzles and read self-help books.

Which made me feel like an ass. It made me feel like a solipsist when the real world suddenly couldn't be ignored any longer. It just hadn't occurred to me, in my selfishness, just how much my presence mattered in that house. I felt terrible then. I thought that I was abandoning her, but I decided instead to believe it was the best thing for her. Maybe she would find the motivation to actually *do* something with herself.

I had to go. I couldn't be a thirty-year-old boy living with his mother, whatever the reason. No matter how one or the other of us managed to justify it. I think I'm going to run off on another tangent here but I think it's as good a place to put it as any.

The way I got started on the Venusian Arts a few years ago, I'd stumbled upon it while researching a theory of mine. Yes, I do have theories from time to time and I *do* do research on them. And one thing that I unearthed after another led me to this, the art of picking up chicks, and it became something that I pursued, not so much to the end I'd originally wanted but more of a deviation or a distraction.

But the whole point was my amazement at how badly Western civilization was failing. How many men had midlife crises, how many marriages resulted in divorce, and screwed-up children who grew up to perpetuate the cycle, and how democratic society as a whole was corrupt and self-destructive. That's what led me to realize that many of these people, myself included, never entered adulthood. Sure, we all aged past twenty-one, but without clearly-defined rites of passage, that didn't amount to much.

Think about it. The Spartans had their agoge, we have summer camp. The crocodile people in Africa undergo ritual scarring in their adolescence, and plenty of other barbaric cultures have similar ordeals that mark the passage of a boy into manhood. What do *we* have? Senior prom and college fraternity hazing ceremonies. Those aren't initiations. Those are photo album memories to be remembered fondly or laughed over later while drinking beer in the dorm. The result is obvious.

And then what do we have to further prevent the transition? Gender confusion with men being taught femininity and the "men's movement" that allows wimps to cry together. It makes me sick. Here we are, the result of billions of years of evolution, the top of the food chain, and we would make our ancestors puke in their soup at even a reality show commercial. I see "liberated women" imitating men's worst traits, like unprovoked aggression and machismo, and sensitive men acting like princesses with their daintiness and easily bruised egos. And I see very confused people stuck in between.

So, I started looking into what makes someone a man instead of just a guy, because to be quite honest, I was in the same boat as all these other jackasses. The only difference was that I knew it.

Also, the girls I'd pull, we never lasted more than a few weeks, because they were always just *girls,* not women. They're always so charming at first, and each one was the *one* until they all turned out to be just another girl who got too many hugs.

Yunno what I mean? She was told her whole life that she was special just for being her, and so she grew up believing it. *I'm Amy and I'm awesome cuz my Daddy says so.* Without being very special at all. Just being rather average, but expecting praise for it. I don't know. The gender roles are very confused in the US.

While I'm not trying to be a sexist at all, I know that as a man it is expected of me to provide and protect. Men that can *not* do those two things tend not to get laid. But the more girls get farther and farther away from their gender responsibilities, the longer they stay girls instead of becoming women. I think. Maybe I'm wrong. But the girls I sleep with want me to pay for dinner, but won't make me breakfast.

Women, for thousands of years, used to be tending their babies and twirling flax and feeding fires, actually doing something worth doing, but now they're automatically entitled to diamond rings, held doors, free dinners on every date, and if not Prince Charming, at least Santa Claus.

Now, I managed to get sidetracked by the other stuff I came across in my research and continued to be a guy instead of a man, just a guy who got laid a lot, which isn't the same thing, no matter who says it is. And that ended up just making it worse.

But, anyway, the original motivation was stirring inside me again and I had to get away. Away from Mommy's house and the excuses we both made for why I was there. Away from the bars and the clubs and the dead end I was hurtling toward at breakneck speed.

So, Elizabeth and I made our plans, and she left on the southbound train back to Miami International. I found myself watching the palm trees that grew out of the sidewalk, in their patches of white-painted pebbles as I drove back. And I found myself smelling the salty air. I suppose I was making conscious memories just in case I never came back. And while I was memorizing everything that I had always taken for granted, I realized that I wasn't going to miss this place at all.

VII

Now, before I go any further, I'm going to take a moment to tell you about my Dad.

One thing my Dad likes to do is humiliate me in front of people, and his favorite thing to humiliate me about is my taste in women. No matter how hot a girl I kiss may be, he will tell other people she is a dog. It doesn't matter who she is. And he will tell her too, to her face, that every girl I kiss is hideous. And he will smile his shit-eating smile at the look on her face.

That's just a little background before I get to the story about him and Mom, to set up the punchline.

Long ago, when we lived in Ireland (and that was my parents, my younger brother and me) my younger brother brought a friend of his around, a girl from Belfast. She looked rather like a potato, and not just because of her shape, but the freckles. Freckles are fine, I think, and looking like a potato is fine, too, if you have that inner beauty that shines through, but this girl was an evil bitch so in this case it wasn't fine.

She came on to me, but the interest wasn't mutual, and since I didn't want to rub my brother the wrong way by insulting his friend, I told her privately that we could not date because she was his friend and he had made it very very clear that I was to keep my hands off of his friends. So, I'm sorry, but our stars're crossed. She seemed to accept that, until my cousins came to visit (remember my cousins, from the beginning of the story?) and they were quite taken with her because of her being an opera singer (like my brother) and, when they heard she spoke fluent Italian, they told her all about the trip we were going to take together for two weeks throughout Italy and the Med.

And they invited her along.

She looked at me, and a cunning look came into her malevolent pig-like eyes.

"But Evan's not coming, too?" she asked. "Cause he's singing all month."

"Well," my cousin Eric said. "He'll just be with us in Corsica and Sardinia."

And yunno when, in the cartoons, the bad guy's eye gets this gleam? A tiny pinpoint of light goes *ding!* and he smiles evilly? Yeah, it looked exactly like that.

So, my Mom decided not to come along this time because she and Dad were fighting over something or other, and we traveled through Italy and had a blast up until arriving in Civitavecchia on the coast, where Dad's boat was docked.

And there she was.

Wearing these white Capri pants and dock shoes and a super super tight shirt with a navy blue anchor on it, and bigass sunglasses, and a little captain's hat with another little anchor on it, this time in gold.

My Dad was surprised, but my cousins were de*lighted* to see her and explained to Dad that they had invited her along. Hope you don't mind, Uncle Ron. It began that very day, the wooing of me by her. If I was below in my bunk, she'd come and try to talk to me, if I was coming through the companionway to get up to the salon, she'd position her massive bulk in my way so I would have to squeeze past and rub my manhood against her. She'd try to kiss me.

At first I was polite.

But eventually I had to explain to her that it just wasn't going to happen, no way, uh-unh, never. I was as easy on her as I could've been. Naturally, she felt rejected, and since I was hanging out with my brother and my cousins all the time, she spent time with my Dad.

They were always walking fifty paces ahead of us and sitting next to each other wherever we went, and I was happy because she was not hanging around *me*. Until that night when we were in Monaco.

I have to call a quick time-out because this is so funny I have to throw it in. We were in a discotheque in Monaco, the kind of club where I spent three hundred frickin Euro on six drinks and wasn't even tipsy, and I made out with this smoking hot Chinese orthodontist chick that was there for the dental conference and I did it by telling her dorky boyfriend that my father was the guy that had invented the brace. He was fawning all over my poor Dad, who hadn't a clue what was going on, and Dr. Juleng and I made it very clear to the Potato what kind of women were my type.

Later, when the boyfriend came back all disappointed and took Juleng out of my life forever, I went back to the dance floor with my older brother and our cousins. While Madonna's new hit single was playing, these two blonde chicks came over to dance with me. I couldn't believe my luck. They both seemed really into me and were making me look very good for the MySpace photos Eric was surreptitiously taking.

At least until.

One of them leaned in and shouted into my ear, in a French accent, that they wanted to spend the night with me in my hotel.

Without even asking my name.

Hmmm.

"Um, I'm not staying in a hotel," I said politely. "I am staying on my Dad's boat."

"Oh."

Pause.

"We would like to spend the night with you on your Dad's boat, then."

I had to laugh, and then I remembered that little bit of French that I'd learned from watching Eddie Izzard and putting the subtitles on, and was so happy I had an opportunity to use it.

I said *"Qu'est-ce que vous dit? Sexe pour l'argent? Vous êtes une putain!"*

And her eyes slid off to the side for a second as she did a facial shrug and that yes-no tilting of her head, before saying:

"Um...Oui."

So, I think that was funny, and if you don't understand, you can Google it.

Anyway, we were back on the boat in the cockpit all chilling with nightcaps and chitchatting, and my Dad was sitting in his deck chair and the Potato was in hers next to but kind of catty-corner to him. Sitting there with her eyes like a toad's regarding me coldly.

Dad was talking—no surprise there, he's always talking; in fact, he *hates* it when someone else is talking—and he was in the middle of telling us something or other about the Battle of the Dardanelles (no joke, he was *always* redirecting a conversation to the topic of war) and she slipped off her dock shoes and lifted her tree-trunk legs to put her swollen feet in his lap.

And while all the rest of our eyebrows jumped he laid a tender hand on one of her feet and stroked it as if it were the most natural thing in the world.

Skip ahead skip ahead skip ahead.

Dad broke up with Mom, but didn't get divorced until many years later. While they were separated and Dad was living with the Potato in Sardinia, he and his mistress found themselves with child. Now, this is one of those lesser-of-two-evils choices my Dad had to make, and he managed to choose a third and downright ridiculous greater evil. Rather than give birth to an illegitimate baby, Dad coerced my younger brother into marrying his friend. Their spawn would grow up knowing his own father as the father of his mother's husband. And, despite my Dad's angry condemnation of that movie "Green Card" because of its moral turpitude (my Dad would condemn anything that was not very *very* moral) he tried to use this awful marriage to get a green card for the Potato. Keep in mind this was the strict disciplinarian who insisted on always doing the right thing "because it was the right thing to do."

My brother was bribed handsomely with promises that were never kept, like a nice car and a nicer apartment, and one of his conditions was that their wedding be kept a deep dark secret. Dad agreed, and then publicized it all over their city.

He invited *everyone* they knew, even the dockmaster of the marina, and embarrassed my poor brother. The Sardinians were all very happy for the young couple, and then horrified to see Dad walking around holding hands with the bride. He lost the respect of everyone they knew, but now blames his lack of friends on the treachery and fickleness of inferior people.

So here's the funny part, at least to me: that old bastard, who was disgusted by every beautiful woman I kissed, had a baby with someone who wasn't good enough for me.

Anyway, the reason I bring this up?

He called from my grandmother's, his mother's house on Palm Beach, inviting me and my new girl to dinner at Taboo, a very formal restaurant. He'd heard through the grapevine that I was going to Guatemala, leaving my mother, and wanted to congratulate me.

This was right before Elizabeth left. I want to squeeze it in right quick before I get to the weird part.

So we go to Taboo, Elizabeth and I, my grandmother, Eric, my Dad and the Potato. I got to wear the nicest of my suits and Elizabeth went out and bought some really nice clothes and new shoes to wear. Well, actually, I bought them because there was a problem with her credit card, but she assured me she would pay me back when I got to my new home in Antigua, so that's pretty much the same thing, right?

So we went, and dinner was lovely except for the times when the Potato burst out laughing in her shrill soprano voice, and everyone in the restaurant except for my Dad flinched and looked daggers at her.

Elizabeth was as charming as ever, and I knew there was nothing Dad could say against her, so I relaxed.

At the end of the night, when Dad had paid the check and we were thanking him, Elizabeth excused herself and Dad gave me his full attention for the first time.

"She's putting on the dog," he said.

Ever since moving to Ireland, he embraced all of the slang right down to calling a wrench a spanner. I didn't ask for an explanation, but he gave it anyway.

"She is not who you think she is. It's all a show."

I made a tight-lipped nod, telling myself he was jealous and doing what he regularly did, and I should take it with half a grain of salt.

"But go with her," he continued, surprising me.

The Potato was nodding her encouragement, all enthusiastic about me leaving my mother behind.

"Go," Dad said, passing me an envelope that I'd later find to have three grand in it. "Have yourself an adventure. See the world, fall in love. But keep your wits about you."

I nodded, and looked to see Elizabeth coming back in her brand new pewter heels, looking radiant as ever.

Now comes the weird part.

So here I am uprooting myself and not planning to look back, ever, and in the two weeks before I am set to leave, I fall in love. Well, sort of.

It was so bizarre, that out of the blue, this cute woman I barely know on MySpace decides she wants a divorce, and she contacts me. She—get this—wants to consummate her divorce by sleeping with someone else, and she just scrolled down her friends list and picked me. Stranger than fiction.

It occurred to me it might be Elizabeth testing me by picking someone on my friends list and asking her, as a fellow woman, to see if I'd cheat. I imagine there are a lot of women in that big pink mafia who'd help a stranger do that at the drop of a hat. So, I flirted back, but I kept my distance, and I guess she liked that. Anyway, long story short, she and I made plans to come over and "watch a movie" at her house over in Tampa, three or four hours' drive away.

The day I was to go over there, this girl I know, an artist, came by to pick up the matt board cutter she'd lent me, and I told her I had to hurry. She asked why, and we were friends enough for me to tell her. She gave me this look and said she hoped I wasn't too dense to get what she was hinting at, but I didn't have to go *all the way* to Tampa because—and she made this downward sweeping gesture with both hands to indicate her own body.

I couldn't believe my eyes. But, a date is a date.

"I'd never stand a woman up," I told her. "I couldn't be that guy, you know?"

"But do you understand what I'm saying?"

I realized she was going to be really offended if I didn't take her up on her offer, so I decided to be late for my date in Tampa. And *Jesus,* she was a wild one. When we were done, she gave me a playful squeeze down there and said she hoped I had enough left over for Whatshername.

I did, as it turned out.

I'd been a little worried about that, but there was no problem at all. I had taken a quick shower, hopped in the car, and high-tailed it to what turned out to be an army wives' neighborhood, calling Sherilyn—that's the divorcing woman's name—to tell her traffic was a bitch and I was running late.

I got there, and we were both very polite to each other, very well-mannered, me trying not to fall into the pink mafia trap that might or might not have been laid, and her reconciling herself with whatever last-minute value system she'd conjured up before going through with what she had started. We beat around the bush a bit, made some small talk, while gradually edging closer and closer, until we were so obviously inside each other's personal space that there was no point in making a pretense anymore, when she suddenly excused herself and went off to the bathroom.

That door shutting hit me with this finality like a nail hammered in with one blow, and I felt like an ass for allowing myself to get caught, especially when I knew it was coming. I considered making my escape.

Until Sherilyn came back out a moment later, with lacy white lingerie and Giorgio Armani perfume, and there really was no more confusion. This was not someone Elizabeth had recruited to test me. It was my lucky day.

So I get back to Palm Beach, wanting to brag to somebody—and not because I'm a jerk, really, just because I'm a human being and that was awesome, and you can judge me all you want, but you'd want to brag too—so I called Josh. He pre-empted me though, with an invitation to his sister's birthday party at Dr. Feelgood's that night. I said Sure, what should I bring her? And he said Just yerself, champ. .

I showed up looking all sharp with a present I'd picked out after a bit of stalking. I don't know why they call it stalking these days, putting that negative connotation on it, since really it's just doing a bit of research. I'd looked at Josh's profile on MySpace—yes, this was back in the days of MySpace, get over it—and found out who his sister was, then looked at her profile to see what she was interested in and what would be a decent present for her. I couldn't just get her a make-up kit or perfume that she might not like and probably not need.

It turned out she was really into camping, so I got her one of those laser perimeter guards that would wake her up when a large animal came too close to her tent. A Florida panther sighting is nice and all—from a distance. I can't imagine wanting to wake up to one of them brushing their whiskers on your cheek. It was expensive, the perimeter guard set, but I'd just sold several paintings, plus the portrait, plus Dad's gift, so I guess I was in a generous mood.

I put it in one of those boxes I'd ordered off that website The Onion. They used to have these phony gift boxes you could get that would make people think you'd gotten them a membership to the Salt Of The Month Club, and this was their first salt of the collection, from the Dead Sea or Venice or someplace, and they would smile and say Thank you and wonder what the fuck was going through your head, until they opened it and found the real gift and everybody laughed like idiots. I got a bunch of those boxes to give at parties so that, if conversation ever started to flag, there would be something to talk about. Oh hey, look what it says here.

I'm glad I did, because there was this one lady—I know saying lady makes her sound old, but she wasn't, she was just older than me and seemed a lot more mature—there was this one lady who was asking me all about it. I told her how I'd gotten one from my older brother that was for a USB toaster. You plug it into your laptop and it'll make a slice of toast in just twenty minutes. But *that's not all!*

She laughed. She laughed at everything I said, come to think of it. She looked familiar, but I'm not sure where from. I figured maybe she had used to be in television or something, but retired because of the kids, or something like that. Either way, she was the kind of good-looking lady that used to be hot when she was younger, but now was a good-looking lady.

We talked and laughed for a bit, and then Josh's cousin Derek and Derek's best friend J-Mart started carrying on, as they do. J-Mart's name is John Martin but he prefers to be called J-Mart, just so you know. Well, the two of them had gone off to a cowboy bar to learn line-dancing a while back, and practiced it at home until they were good at it, and they do it out on the floor at non-cowboy clubs. At first, other people are always like, what the fuck...? But then a couple of girls go to stand next to them and learn to do it, and then more, and next thing you know, the cool people are all out on the floor line-dancing and all the people that want to be considered cool but are really just assholes stand around the floor shaking their heads at them. And of course, there is no cowboy music playing, just the regular club music, DJ Skribble up there doing his remix of "Hit the Road, Jack" at the moment, so there's this juxtaposition of genres that's fun to watch, and there's Derek and J-Mart being all serious and all the cool cats in the place watching them and thinking Goddammit, those boys are going home with the chicks we've been trying to score, and all they had to do was learn a goofy dance.

Then the song ends, and all the hottest girls come up to Derek and J-Mart, who both have little pocket notepads and pens in their jeans. The girls try to chat, but both boys do the

Time-Constraint move that allows for a brief conversation but prevents anti-climax. They say, Thank you, but I'm sorry, it's my friend's birthday and I really can't ditch her to go off with you, but let my friend take your number. Then they both holler the other one's name, and they switch places, whip out the notepads and pens, and with much authority, take down the other guy's girl's number, then leave.

I figure they do this because they really are not in their element, so instead of accept that it isn't their scene, they capitalize on not fitting in and still manage to make themselves the center of attention. You know, that counter-intuitive response again. And then they make their escape, with the phone numbers, and make plans to meet up later at quieter venues where proper dancing is not required, and conversation *is*, and then they retire to their homes to show the girls how to line-dance in private, and the physical interaction leads to the, um, well, ripening of their relationship.

I had to hand it to them.

So anyway, we all got drunk.

I wasn't hitting on anyone in particular, I think because I didn't feel I needed to after the day I'd had. It felt nice to not have that agenda, and just be myself and have a good time.

I decided to drink tequila. Now, I dunno if I mentioned this before, when I was explaining about Ginger, but tequila does not make me into a psycho like it does her. It makes me all lovey-dovey and romantic, with whoever is nearby.

While doing shots, Josh leaned into me and said Yunno, Jen is not bad-looking for a bigger woman.

I said "Who's Jen?"

"The one you've been talking to, the ex-model."

"Ex-model? Which one? Everyone around here claims to be an ex-model."

"You never saw her in any of those surf gear magazines? Or in Christian Dior with her hair wet and her eyes narrowed and her lips parted in the throes of passion over this bottle of perfume?"

"Maybe I did."

"Well, that's her. Jen, the one you were talking to. The tall redhead."

"Oh. Okay. What about her?"

"Nothing. I just think she looks pretty good for a bigger woman."

"She's not a bigger woman. She's just tall."

"That's what I mean."

"Josh, we're both short guys. Everybody is taller than us, pretty much. But, since you mention it…"

"Yeah, she looks pretty good, doesn't she?"

I did my last shot and got all dramatic, which I supposed happens to everybody. I made this big show of being heroic, like the guy that says "I'm going after that truck" or "Take him away, officer."

"Ex*cuse* me," I said, and forged my way bravely toward the woman who was, apparently, called Jen, just in time for her to get jostled while drinking and have her Bacardi Limon with a splash of Sprite spill down her chin and onto her ample bosom. Poor girl. It wasn't her fault at all, and she was so embarrassed. She looked around for a napkin, but there weren't any, so—keeping in mind I just drank tequila—I said Here. Leaned in and stood up on my tippy-toes to kiss the wet spot on her throat and, well, lick the booze off of her. Yeah, I know, but I was drunk. On tequila.

She let me do it. I could see out of the corner of my eye that other people were watching and going "Whoa, hey, look at this." And some of the girls in our group approved with enthusiasm and two of them didn't. At all. But anyway, this lady Jen let me kiss her throat and her upper chest for a minute and took my face in her hands, guided my lips to hers—yeah, I had to strain a little to reach, because man, she was tall—and we kissed for I don't know how long.

Long enough for Josh's sister to interrupt and suggest we get a room. Jen seemed embarrassed, like she'd lost herself in the kiss and forgotten we were in public. But there was also a look in her eyes I'd seen before, the look of someone who'd once been hot and much sought-after, but who'd maybe gotten married and had a couple of kids. Someone who hadn't made out with a guy in a bar in a very long time, someone who hadn't felt wanted. And she was feeling it now, and man, she'd missed it.

She looked at me.

"Who are you?" she asked.

"Just some guy."

"I'm married. I have two kids, the elder's fifteen. So I'm not exactly—"

"What you are is married to someone who doesn't know what he's got. Some jackass who doesn't look at you the way you deserve to be looked at and treat you the way you deserve to be treated."

I glanced over at Josh, who was grinning and shaking his head at me.

"I have to go home," Jen said.

"Okay." I waited. She said good-bye to everyone else and Happy Birthday to Josh's sister, and then asked me to walk her to her car, make sure she got there safe. We got out into the one-in-the-morning of Clematis Street, where a trolley car was picking up drunk shiny people and taking them to City Place, and bringing drunk City Place people here. We started walking, and I slipped my hand into hers. She was startled, and afraid like her husband would somehow see, but I curled my fingers between hers and walked with her down the sidewalk, pushing through people, leading her through crowds.

Her warm hand tightened around mine. At one point she told me we had to let go because all the friends of her asshole husband hung out at the bar we were about to pass. I didn't want to get her into trouble, so I let her go and even dropped a few steps behind her until she led me to the parking garage. We went all the way up to the top, where no fluorescent lights lit up the inside of every car like broad daylight, but under the night sky in the wind.

What we did wasn't tacky, or cheap in any way. It was romantic, like two teenagers at Lookout Point in the movies, stealing a beautiful moment together and being in love.

Jen gave me her number and I called her the next day. We met for afternoon drinks at O'Shea's, that place across the street from Respect's, and she looked fantastic.

"First of all," she said when we had our drinks and were seated outside in the back. "What happened last night, I don't do that. Ever."

"Me neither."

"I'm serious. I don't just meet guys in bars and sleep with them in the back of my car like two hours later. That's not something I do." Maybe she was telling herself that more than she was telling me. Trying to salvage something.

"I had a wonderful time," I said.

"Me too."

We looked at each other over the rims of our glasses while we drank.

"How old are you?" she asked.

"Thirty."

"You don't look thirty."

"There's a painting of me that ages instead."

She looked at me a second, then smiled. "Oh, right, Dorian. Good one." She drank again, just to be doing something, then: "So what do you do?"

"I'm an artist."

"Really? What kind, abstract?"

"Oh, God no. Funny you should ask, though. I just put a manifesto on a couple online forums—well, another time." I didn't want to go off on a tangent and get into one of those friendship-ending What is Art conversations with her. Christ, that's even more dangerous than a political or theological discussion, especially on a first date.

"I do all kinds of things, really. Mostly realism or surrealism, but everything from portraiture to wildlife and still-life to nudes."

That last one always piques women's interest, but most of them always let it go unremarked-upon.

"And what about you?" I asked.

"This could never work. I'm married. I've been married for seventeen years. I've got two kids—"

"You look great in that outfit."

"Really?"

"It probably takes a lot of self-discipline to have great skin like yours, especially in South Florida. You definitely take care of yourself, and you really don't look like you've had two kids, much less been married for seventeen years. If that were true, you would have to have been married at, what? Ten? Eleven years old? But you weren't. So you've got to at least be thirty-four, but honestly, you don't look a day over twenty-seven, twenty-eight, tops."

I drank, just to be doing something.

"Thanks," she said, appreciating it, but knowing it was a line. "Thanks for the compliment."

"Oh, it wasn't a compliment."

"What?"

"I wasn't trying to compliment you at all. I was just being honest."

The look on her face as I glanced at her over the rim of my glass said Home Run.

We ended up going back to her place because her kids were in school and her asshole husband was at work, and he wasn't going to be home anytime soon because he always stopped at the bar afterwards "to unwind." She gave me the grand tour of her neo-something house, and we stopped at her daughter's room for a moment, where I was blown away, not just by the posters on the wall but the level of admiration the teenage girl must have had for her Mom. There were these surf-gear posters of a knockout redhead looking wet and passionate on the beach, and others of her about to orgasm over bottles of perfume, and others where she looked oh-so-serious in her haute couture ensembles.

I knew this was the moment to say anything but Wow, that was *you*? Or the equally-stupid You look even better now, so I just skipped it all and held her hand again, tugged on it gently so that she moved toward me with a faint smile, and we kissed. She backed up while we were kissing tenderly, and I followed, and we kissed our way up the stairs, down the hall, and into the master bedroom, where we made love softly for hours. And when I say we made love, I mean every touch was with perfect tenderness and respect and care. She cried quietly when she said it was time to go, and shushed me when I tried to protest. Ushered me out the door and told me to call her tomorrow.

I drove away feeling very sad.

"Listen," Elizabeth said in the email I read later. "When you get here, there are a couple of things you should know."

It felt so strange to be tearing my mind out of the new reality I'd formed with Jen, especially on top of the lucky day I'd had, to be reading my actual girlfriend's words. I'm sorry, but it did. It felt like something from someone else's life intruding. I think it was because of all of the weed she and I had smoked, those days were like a half-remembered dream.

I did feel guilty, but it felt more surreal than anything. It took a moment for me to transplant myself once again in the new direction my life had taken, extracting it from the even-newer one.

"If Toto's in Antigua, you need to avoid him at all costs. If he finds you in a bar, he'll come up to you and be as friendly as he can be. He'll become your new best buddy, and you'll realize he's not all that bad, and he'll start offering you help with this and that, the kind of things you need help with in a new city where people speak a different language than you do. And he'll do everything you need out of the goodness of his heart. You will let your guard down.

"And then he'll pull out one of the many guns he's got hidden about his person, and stick one of them in your face. He'll make you beg for your life—"

I couldn't help but think of Terry Griffin here.

"—for a little while, the length of time depending on how big of an audience he has, and then he'll tell you that you've got twenty-four hours to get out of Guate, or you're dead. And he means it. And then he will come find me wherever I'm hiding and give me an ass-kicking, then go get piss-drunk at Café No Se. I can imagine that look in your eye, I bet you have it right now, don't you? But he has got so many of the police in his pocket, either because they're his cousins, or his customers, or just plain friends, that he can get away with firing his pistols drunkenly into the air in Parque Central, in broad daylight, with no regard for where those bullets come down, and has never gone to jail. So you can imagine what'll happen if somebody was to off him. If I could get my hands on his phone, though, I could get his contacts and take over his business, and then find a way to bury him."

I thought I remembered those words, from a film I had seen before. They seemed so familiar, and that eerie feeling of *déjà vu* made me wonder if I'd been in this situation in a past life. Or if this happens so often, damsels in distress appealing to that desire all men have to be their hero and either defend them or avenge them against evil, conjuring up a savior by telling some dumb guy about her woes and waiting for him to assert himself, if it happens so often that it's wired into our brains to see it coming, should I know I am being manipulated and cut my losses and run, or be the brave man I am expected to be? Be the hero, or the sucker?

And was I just over-analyzing this?

Where did I read or hear those very same words before, practically verbatim? Or is it just my imagination? Just *déjà vu*, which happens sometimes and no one knows why?

Whatever the reason for it, my spidey-sense was tingling and I knew something was wrong, but for the first time in a *long* time I knew I was going to be on a real adventure, with genuine chases and escapes and opportunities to be brave and heroic. It felt kind of like going off to Mississippi with Ginger, feeling a bright future ahead of me, the possibilities wide open.

The rest of Elizabeth's email was just chit-chat about her day-spa and local gossip she felt I needed to brush up on so that I wasn't completely out of the loop when I arrived. She wanted to see that look of Ahhh! when she introduced me to someone, the look of a person with prior knowledge putting a face to a name and having contexts match up in his mind.

She told me about John Rexer and Michael Tallon from Café No Se, which was the self-appointed dive bar a couple doors down from the day spa, and Mono Loco, which seemed either to be a themed franchise sports bar or would be the first of a chain eventually. Reilly's was the obligatory Irish pub that every tourist city in the world had to have. Funny that this one was owned by a Dane and a Welshman.

Across the street from Reilly's was the Casbah, the only discotheque in the city, and both were on Calle del Arco, Fifth Avenue with the famous yellow arch everyone took pictures of and all the street artists painted watercolors of with the picturesque Volcan de Agua looming in the background.

A short walk further down was the big yellow church La Merced. Around the corner was Hector's, the restaurant with no name and nothing to identify it from outside. Hector himself called it the Restaurant, and everyone else called it Hector's. Elizabeth said she had loved it right up until one of her friends found broken glass in his bœuf bourguignon. The guy was eating with friends when he bit down on something hard, and in an effort to be polite, decided to just swallow it and not make a big deal, but at the last moment coughed it out.

Instinct, or God, maybe, didn't allow him to ingest what turned out to be a shiny pebble of glass. Horrified, he dug through the rest of his meal with the tines of his fork and found a *lot* of it, little shards that could not have been there by accident. He sent it back, and it still appeared on his bill. That had happened just the other night, and they were telling *everybody* about it, so she was telling me too.

I wrote back that it was stuff like that that really pissed me off. I see people singing about how we have to give peace a chance, and blah blah blah, but there would never be peace in the world until people stopped messing with other people's food in restaurants, pissing all over public toilet seats, spray-painting public property, and just generally putting their mark on what belonged to someone else.

Like "peace" is something that men in government prevent on purpose, but the common man could be a part of if only we all held hands. I know I went off on a rant in my response, but it's something I've always felt strongly about.

Then she replied, agreeing with me, and saying something about men fucking other men's wives.

That made me stop.

VIII

Jen called and told me I had to meet her at O'Shea's. I asked why and she said Just be there. So, I showered as quickly as I could, brushed my teeth and got dressed as nicely as time would allow, and rushed off to Clematis. Parking was a bitch, but it always is. I felt like a jerk hurrying down a South Florida sidewalk and starting to sweat after going to the trouble to be so clean, and I also remembered with a bit of chagrin that no woman wants a man that'd do exactly what I was doing, so I slowed down and walked the way a man should walk.

I realize that if Cliff's Notes ever makes a version of this book it would probably say this moment marks the admission that I am still not a man, despite all of my efforts to artificially appear one, blah blah blah.

Bullshit.

I was just excited to see her, especially at such short notice, so Cliff can shove his psych evaluation.

I got there and Jen was out back, in her seat at our table, with her drink in her hand and one waiting for me across from her. Nice touch. There was even another glass next to it, with just ice and a little bit of water. Even nicer touch. Wow, she really liked me.

"I had them leave the ice in a different glass," she said. "So it wouldn't melt and water your drink down before you got here." I gave her a long, tender kiss before taking my seat and putting some of the ice in my glass.

"So, how are you?" she asked, stalling.
"I'm good, you?"
"Fantastic."
"Oh?"
"I left my husband."

My hand stopped just before the rim of my glass was at my lips, and I froze with my mouth open, eyes wide and staring at her. She smiled, and then all the trembling, pent-up energy came out of her, everything she'd bottled up for all these years.

"I didn't do what all my friends told me to do, you know, change the locks and leave all of his stuff outside the house in boxes because I think that's the kind of thing tacky, cowardly people do. I just told him this morning when we were having our coffee. He was ignoring me, like he usually does. You know, not ignoring me as in he's oblivious of me being there, but ignoring me like he wants me to notice he's ignoring me and feel like shit about it, which I have for years. Can you believe it? He got *mad!* I mean, can you *believe* it? The last thing he'd said to me was the day before, while I was making dinner. Making him *his* dinner, and what does he come in and say before going to plop down on the couch and watch the game? He asks when I'm going to take my fat ass to the gym. I mean, can you *believe* it? And then he has the nerve to get mad at me today because I tell him I'm leaving him. He started yelling at me, accusing me of cheating on him, and I felt more calm than I think I ever have in my life. I told him No, I wasn't cheating. I was just fed up with him and the way he treated me, and I wasn't taking it anymore.

"Then he said that he knew I was cheating on him because he had me followed the other night and he saw me kissing someone. I said Really. You really had me followed, and he said Yeah, and I know you kissed somebody. And I looked him in the eye and said No, you didn't have me followed, because if you *did* you'd be accusing me of a hell of a lot more.

"Then I called my Mom and my two sisters and packed my stuff and moved out."

She sat there, across from me, with this ever-so-pleased triumphant look on her face, and I was nodding and smiling and thinking *What have I done?*

I called Josh from the bathroom and told him all about it as quickly as I could, and there was a long pause on the other end of the phone. A long, pregnant pause.

"Hello?" I asked.

"You're going to Guatemala?" he asked.

"Yes. I'm leaving soon," I said quickly, impatient that he was still processing that.

"You're leaving."

"Yes!"

There was another pause, and I knew I'd run out of time and had to get back to the table or Jen would know I wasn't just washing my hands, so I said I'd be around later and hung up. I took a deep breath, then saw myself in the mirror, the harsh accusing eyes my reflection was looking at me with.

"Yeah," I mumbled. "I know."

I went back out and slid into Jen's seat with her, wriggling my knees underneath her thighs so that she could sit on my lap and put her arms around me. I needed to be gently kissing her and caressing her warm cheek with my own so the pressure of saying something would be postponed at least for a moment. She seemed so happy, and the responsible thing to do would be to crush her happiness, but I couldn't bring myself to. I had to keep her as far away from pain as I could, for as long as I could.

Then I started thinking that I shouldn't leave at all, that I should stay here with this beautiful woman and make things work. Stop going to bars all the time and get a proper job so that I could support her and face the consequences of what I'd done. But then, there was Elizabeth, waiting for me in Antigua. Would I break her heart instead? And what was I *doing?*

Could I go off to Antigua and leave this woman behind? Could I love two women at the same time? I knew Elizabeth's disappointment would not last forever, and of course, neither would Jen's, but Jen had made a monstrous change in her life because of me and wouldn't have me around to make it easier.

She would feel like Wile E Coyote when he realized the ground wasn't underneath him anymore, and I couldn't make myself feel any better about it by saying Hey, either way, she's better off without that asshole.

And Elizabeth was counting on me to show up and be her knight in shining armor, saving her from her psycho ex. It would look like I had chickened out, and that was even worse than falling in love with somebody else.

I held Jen tighter, and she melted into my arms.

I held her like I'd never let her go.

I went home and started deciding what I'd take with me, if I decided to go, and what I would leave behind. If I was just going for three to six months, which is all I could legally do, I could come back for some things, and if I stayed forever I could part with lots of it. My collection of first editions were a hard decision to make. When I finally decided to leave them, I took some out of the bookshelf and opened them to random time-yellowed pages, held them to my nose and smelled that delightful fragrance of Literature that I knew I'd miss.

I went through all of my drawers and regarded various things I found there. My old name tag from when I worked at ER Bradley's Saloon. There'd been the legit one that just said Johnny, and then this one, the one I'd swap the real one for at ten o' clock when my boss left and I was in charge. It said El Cunto, and you wouldn't believe the hit it was with customers. They would laaaaaugh, and I invariably kissed at least some of the cuter ones when the restaurant became a club at ten-thirty.

People're funny like that.

I stared at it for a while, grinning, then put it back.

I had a bunch of half-finished paintings, all of them of a different woman reclining nude in the odalisque pose. Yunno, that pose where a woman's lying on her side, propped up on throw pillows? She usually has one arm up so she can lean on them while her body curves gracefully, with one leg bent and foreshortened while the other one's fully extended? That's my go-to seduction, but so often, I never finish the paintings because I'd never spend more than a few days with the girls who posed.

Lemme explain. There's this thing you can do when you make a date. Normally, guys'll take a girl on the same paint-by-numbers date—pick her up, take her to a restaurant, chat for a while, take her home, have that awkward moment where you maybe kiss, maybe not, maybe get invited inside. The whole time, she is in control of what will happen later. I say there is nothing special about a date like that, so I change it up a bit.

What I do is ask her to come pick *me* up, because my car is in the shop, I need to have it looked at, whatever. So, she'll get to thinking about the role-reversal of being in the driver's seat. She'll feel more in control because I will be at her mercy, depending on her to bring me home. Then, when she arrives, I answer the door nude except for a towel wrapped around my waist. She stares, with wide eyes, at my washboard abs as I say, "Oh, hey, sorry, I'm not ready yet. But come in."

Then I stop and make a face, the kind of face girls tend to make when they ask if you have a condom, and ask "Would you mind taking off your shoes?"

I gesture with a tilt of my head to the shoes sitting there by the door, and she always hesitates, then smiles uncertainly, and slips her shoes off. All the while looking at my abs and the towel wrapped around me, wondering if it will fall off, and she comes into my territory, feeling both a little vulnerable and at an advantage at the same time.

"I'll just be a minute," I say. "Want a glass of wine while you wait?"

She makes a yes-no tilting of her head and says "Sure."

I lead her to the kitchen in my towel, letting her feel the cold tiles under her bare feet, letting her feel "at home" in a stranger's house, and feel the tingle of being alone with me in such a familiar and personal way before our first date starts.

I show her the choices of wine, and we both put on silly British accents while we discuss the silky texture and vibrancy and body of this, that, or the other one, and we laugh. We pick one and I make the little man of the corkscrew raise his arms in exultation, then pull the cork out with a loud *pop*. We toast, watch each other over the rims of our glasses while we drink, and then she looks around the house and invariably asks about the paintings. I show her around, telling her we won't go into the guest bedroom because that's where my Mom stays whenever she visits me, and I leave it the way she left it, to not intrude on her privacy. She is usually out playing Bridge or off in the Virgin Islands or at the Golden Door anyway.

We end up in the room I use for my art studio, which is right next to my bedroom. She's always amazed and never fails to give me a sly smile, biting her lower lip, and says she wants me to draw her.

"Really?" I ask.

She nods.

I grab my sketchpad and a pencil, and take her by the hand, feeling the *zing!* of electricity when my fingers curl into hers, and lead her back to the living room. She feels us getting farther and farther away from the bedroom, and relaxes even more. I sit her on the couch, pretending to have forgotten I'm pretty much naked, and my hands are confident as I hold her, arrange her into the pose I want, never apologizing for touching her. Holding her knee and ankle to bend one leg on top of the other, touching her bare foot tenderly to move it to the angle I want, and then sitting down in the chair across from her, the sketchpad across my knees. Still in my towel.

She finds it exhilarating to be looked upon in such a way, to have her body, every line and curve of it from her face to her toes truly *seen* by someone else, and studied, and to surrender to the mixture of embarrassment and excitement. It is a liberating feeling, and it makes her feel tremendous power in her own beauty. Maybe she has always felt attractive, always felt desirable, but now she feels beautiful.

The burn of my gaze when my eyes leave the page to look at her again, she seems to *feel* it caress her face, and the vulnerability of that burn as she watches me gnaw on my bottom lip and squint, my pencil scratching away at the sketch pad, makes her fidget, and smile warmly when I reprimand her with my eyes for moving.

Then, when I take the pad over to the couch and kneel down beside her to show her, the change that comes over her face is like warm light, and I am only inches away for her to take my face in her hands.

We never end up making it to dinner.

The next day, she wants me to paint her, in the nude, and comes over, and we start, but the painting never seems to get finished.

I found so many of these paintings in my closet, and sat there on the floor, remembering each one and reliving every glorious moment. I decided that I would have to leave them behind, since I was never going to finish any of them.

Eventually, I found the Book.

Another friend of my Mom had spotted the budding li'l Casanova in me when I hit seventeen and had given me one of those little black leather books that said 'Blondes, Brunettes & Redheads' in gold letters. I chuckled and opened it, looking at all of the names and numbers, and brief notes scribbled to remind me who was who, like 'Rides horses' or 'Plays violin.' I sat down on the edge of my bed and remembered each one of them with all of the love I'd felt for them at the time. I flashed on the moments we'd spent together and wondered why we'd broken up, and where they were now, until I got up and went to the desktop. I got into MySpace and the new one Facebook to look them up.

Katie wasn't on either of them, which was a bummer. I haven't told you anything about Katie and I'm going to keep it that way. She was special. And I'd been a real jerk.

I went through the book and found most of them, and I sent them all friend requests, until I got to Tamara and paused.

Was that a sleeping dog I really should wake?

How many times had I broken her heart and put it back together only to break it again? Every time I did it, I felt so bad afterwards I couldn't face her, and when Bipolar Johnny was gone and I ran into her again somewhere, I'd apologize profusely, saying she didn't deserve the things I'd said, and then she'd take me back, and the cycle'd begin anew. She was just so innocent, so sweet, that even though she was a little bit older than me, I always felt guilty, like I was taking advantage of a child. And that became self-loathing that would ticktick-ticktick down to the Explosion.

Really, I felt like a pedophile when I'd make love to her, and nothing could erase that horrible feeling. I would burn with the sin of sullying her until it was too much to bear, and then I'd break her heart. Two days later I would wonder what the hell I had been thinking to just throw away a prize like that, why I couldn't just keep my stupid mouth shut and get over whatever it was, and then I'd start pining for her.

Should I call should I call should I call?

At least just to say I was sorry?

She was probably married by now and safe from my evil clutches, as well she should because she was a hell of a catch. She probably had kids. Maybe a little girl with the same wavy Sicilian hair, or a boy with the same disarming smile.

I should at least see.

I should at least call her up and say Hi.

I dialed her number and I waited, listening to the ring with my heart in my mouth until the voice of an older woman answered. It was startling to hear such a mature voice, and I was not sure it was her. I began to stammer and cough, and she asked patiently what she could do for me. I paused. Here was the Rubicon.

I could hang up now or speak out.

Back up into safety or dive off the cliff.

"Um...Tamara?"

Something changed in the silence on the phone.

Her silence went from quizzical to hostile, as if I heard her eyebrows dropping from their arch across her forehead to knot together over hardening eyes.

"Who is this?" she asked.

I swallowed hard. I deserved this. I knew it.

I could still hang up, though.

But then I wouldn't be much of a man, would I?

"It's Johnny. Johnny Yen."

And if you've ever felt like Wile E. Coyote when he realizes the ground isn't beneath him anymore, you'll know what I felt like then. Right when his ears droop and his neck stretches long and thin because his body is plummeting to the ground a thousand miles beneath him, right before his head snaps elastically out of the frame to catch up with it.

That moment of Ohhhh Shit.

"What the hell are *you* calling for, you son of a bitch?"

"Tamara?"

"Tammy was my *daughter.*"

"Well, is she—" and then I realized, a second too late, that she had said Was.

"No, you little shit! She isn't here! She hasn't been here for a very long time! You took my little girl away from us—"

I hung up the phone, shaking like a leaf.

I couldn't breathe. My heart was beating so hard that my vision was crimping at the edges, and I jumped half out of my own skin when the phone rang in my hand. The same old ring I'd heard a thousand times before became somehow as harsh and accusing as the voice I knew I'd hear if I dared to answer.

Damn you, Star 69.

I could never answer this phone again. I stared in horror at it, unable to move.

"You going to answer that?" my Mom called.

Tear the phone cord out of the wall, I thought.

Whatever you do, don't answer, I thought as I lifted the suddenly-heavy phone to my ear, pressing the glowing green button. I couldn't speak. My mouth was so dry I could only croak. Oh my God. Tamara was dead and it was my fault.

"Johnny!" Josh shouted. I could hear a loud rushing of wind as if he were calling from the bed of a speeding pickup truck, or an airboat skimming across the Everglades.

I gasped and sagged against the wall.

"Listen!" Josh said. "We're throwin' a Bon Voyage party for you at Respect's tomorrow night, so you better be all spiffy in your Sunday best!"

I couldn't believe my ears. One, a party for me would've been a pleasant surprise on any day, because I'd never think I deserved it, and two, I damn sure knew I didn't deserve one now. I tried to reel myself in, pretending that the other phone call that just happened *hadn't* happened, and I was still in this happy world where I didn't know I'd pushed that poor girl too far one night. This other life I'd had up until two minutes ago, when I found out there were consequences to the stupid shit I did. Tammy was dead. Tamara Musso killed herself, because I broke her heart.

I shut my eyes and saw her face, sweetly smiling as she gazed lovingly up at me. I saw the way she lit up whenever I'd tell her I was sorry, and that I loved her.

Saw her notice me in a crowded room when I walked in, and the way she'd wink at me while talking to her friends.

She was the most beautiful girl in the world.

And I deserved to be dead instead of her.

"Awesome," I said weakly. "Thanks, bro."

"You bet! Catch you later!"

I nodded, as if he could hear it, and hung up.

The phone rang again, and my heart turned to stone. I squeezed my eyes shut tight, lifted the phone once more to my ear, and accepted the call.

Silence.

I waited, my breath held, and the silence was worse than what I was expecting, until I heard a very nervous voice asking "Johnny?"

I couldn't believe it.

Of all the things that could have happened, of all the people that could have called me up out of the blue, especially in that dark moment, Katie was the last one I'd expect.

"Katie?" I gasped.

"Um, hi."

We talked for hours. She had called me from Connecticut, having just broken up with her asshole boyfriend, whom she'd confronted with his *other* girlfriend, and kicked him in the nuts, then came back to her place shaking with rage, looking for someone to talk to. For some reason, she looked me up on MySpace, after all these years, saw that I was having a Goodbye party, and four-elevened my number.

Now she was throwing caution to the wind, saying Wherever you're going, don't go. Come up to see me instead, we'll go to Vermont where my family has a Swiss chalet, and...you get the picture.

And I wanted to. With all my heart, I wanted to. I would have left all this mess behind me and gone off to be with the one I really wanted all along, but seeing myself as I really was only a moment before, I refused to put her in that position. Suddenly knowing what kind of a monster I was, I couldn't do it to her. It was a hard choice, and it was hard to tell her No, but it was the best thing for her.

Next night, I went to my Bon Voyage party, really rather chuffed that one had been thrown for me, and wondering if it would be a big turnout. I was still on the fence about whether I should go to Guatemala, or stay here and be a man with Jen, and I figured I'd confess everything to Josh and ask him what he thought.

When Respectables came into view, I squinted at someone in a Joker tee shirt with a two-foot-tall Mohawk, wondering who it was. I mean, it kind of looked like Josh, but...and then I felt like an idiot. It *was* Josh, I had just never seen him without his baseball cap, and with his hair gelled up.

Wow, I thought, how little I pay attention to my friend.

I got to the door, where Josh was leaning against a palm tree with the tall fat website designer in a Decepticons tee shirt and a few other people I didn't recognize. Josh smiled evilly at me, and looking back on it, I totally misread its meaning.

"The man of honor!" he called, and the others looked at me, all of them affecting disinterest. It made my small hairs all stand on end, as if warning me of danger. I slowed my stride a little, uncertain, but the look on Josh's face vanished, replaced by an overly-friendly smile that made me even more wary.

There was an awkward moment when I arrived, the guy in the Decepticons tee shirt looking like he was trying not to say something. His shoulders were up, and he kept shifting his weight from one foot to the other. Josh, too, was tense, and it made his plastic smile so obviously false.

"Ready?" he asked. I frowned, looked from one of them to the other, and back again, and nodded. Josh set his jaw and reached to pull the door open for me. Music and light tumbled out, and I went into Respectables for the last time.

The place was *packed*. That was the first thing I saw, and it made me feel great, all of those people gathered underneath a banner that read "Bon Voyage, Johnny!"

All of them were talking amongst themselves until they noticed me, and instead of shouting something in unison and throwing confetti, they just went silent.

There was a tense moment of confusion, until I realized they were all women.

And every last one of them, I had slept with.

A spotlight came on over me, bathing me in white light and exposing me in front of God and everybody.

I looked at Josh, who grinned bitterly at me and slapped me hard on the back.

"Here are some other people you forgot to say good-bye to," he said. "So, now's your chance."

I stared at him in shock, my *friend,* who would do this to me, and then I read it in his eyes. All of the times he had been my alibi and my wingman and I had just told him I was leaving in passing, as if he didn't matter. I completely forgot to talk to him about it, to ask his opinion, to even *bring it up,* and I guess that must've hurt him more than anything. Jesus, what a selfish asshole I can be.

I swallowed hard, and looked at all the women looking at me. On cue, the DJ lowered the music, and one of the girls came forward with a microphone in her hand. I couldn't even remember her name. She was one of those goth chicks, albeit a pretty one. She smiled wickedly at me, with her purple lipstick that made her look like a drowning victim.

"Well well well," she said. "Lookie what the cat dragged in! The man of the hour, Johnny Yen, himself!"

No applause, not like I expected any.

"Johnny, you've been a naughty boy."

I both saw and felt a dancing glitter, just like whenever my Dad was screaming at me when I was a kid, and the floor wavered sickeningly beneath my feet.

She put on a mocking face of concern.

"Is your phone broken, or something, Johnny? Did you lose *all* of our numbers? How could that be, if we're immortal? Isn't that what you said, we're your immortal beloved, and that since we never fought, we'd all be frozen in your mind as the perfect woman? You painted all of us, didn't you? In the same pose. Did you think we wouldn't ever find out? Did you think we were all that dumb? I guess you did. Well, we are *so* happy to hear that you're leaving, Johnny. We're so happy that you are going far, far away, to never be seen or heard from again."

I scanned the faces in the crowd, and my heart sank. Every last one of them was looking at me with such hate, and I wanted to die, right then and there, when my eyes settled on Jen. Poor, heartbroken Jen. What the hell was *she* doing there? And then I remembered. Josh's sister's party. I bet she was the first one he called.

"But before you go, we want to give you a parting gift."

Shit.

The DJ put on some ominous music that made the girls laugh, and the fog machines billowed out smoke.

"We know you believe in magic, Johnny. You sure told us that, making us all feel special for one hot moment. Well, I believe in *black* magic."

The lights made the smoke turn green, and the ominous music got scarier. The girls were giggling.

"Some of us have been casting spells since we were little, and so we have plenty of practice," she went on. "So Johnny? We curse you. We curse you by the power of three times three to never find love, ever, anywhere you look. We curse you to a life of sadness and loneliness and misery. And worse, we curse you to *think* you have found love with every woman who will break your heart. Every tramp who will suck you dry, you will love her more than life, until she can stab you in the heart and you'll still try to kiss her with your dying breath. And worse, we curse you to find the perfect woman, one who'll love and honor you, and we curse you to not love her back."

I couldn't breathe. Standing there in that white light, all those eyes burning into me, I couldn't breathe at all.

"We curse you to suffer, as we have suffered, the shame and the rejection of being used up and thrown away, of being made to believe in ourselves and then forgotten. And you will suffer for each one of us. You will never be happy, Johnny."

And somehow, I knew that it was true.

The girl whose name I didn't remember handed the mic to someone and crossed the dance floor toward me, her go-go boots clomping heavily. I shuddered with every step she took, until she was right in front of me, eyes cold with contempt. I saw it coming, but made no move to stop it. Her hand wound back and I braced myself. Her drowning victim lips tightened.

The slap resounded, and it stung almost as badly as the thunderous applause from all the girls in the club. She turned her back on me and walked away.

Another girl came.
I could have turned and walked out the door.
I could have raised my hand to stop her.
I could have done anything at all, but I didn't.
She slapped me hard, and I felt my cheek flame red.
One by one, they came, each of them glaring deep into my

eyes for a moment so I would recognize them, before they wound back and struck me. It was almost like an Irish wedding, when the poor bride has to stand in front of everyone and be kissed on the cheek by every guest, even the ones she doesn't know, the party crashers. It was funny that that is what I thought of.

Then, one of them startled me as she approached, when I could make her face out in the dim light. It was my artist friend, the one who'd lent me her matt board cutter. I must've looked at her like *Et tu, Brute?* because she smiled.

There was warmth and pity in her gentle eyes, and when she got to me she reached out and slipped a small, folded note into my hand, and leaned in to kiss me on my burning cheek.

An awful hiss of anger arose from the crowd, and she had to walk past me right out the door and never come here again, or be beaten to a bloody pulp by all of them for being a traitor. Her kiss hurt my skin, but it meant the world to me.

When she was gone, I opened the note and read

I always knew you were no angel. Bye, sweetie.
Frogbells

I frowned, wondering where I'd read that nickname before, looked up at the next girl coming, and my heart stopped.

It was Jen.

This is where, in the movies, the guy says I can explain! But he never does. He just stands there lamely and watches as she calls him every name in the book and storms away. This is where, in somebody else's story, he tries to tell her that he had changed his mind and decided that he wasn't going anywhere.

But she wasn't going to believe a word I said.

Her eyes were white hot slits of rage and betrayal.

Her perfect lips writhed back in a snarl that tried to hold the tears back, and did not succeed.

Her voice cracked.

"When were you going to tell me?"

I closed my eyes, trying to teleport myself to Timbuktu, or turn invisible and slip away quietly. I wished for lightning to strike me dead. I wished to be anywhere except in front of her. But with my eyes closed, I saw Tamara smiling at me, her adoring eyes so close I could see myself in them.

So, I opened them, and Jen's face split into sobs so hard her shoulders heaved and her knees buckled, and I moved to catch her, but stopped myself. I stopped, and stood as straight as I could. I don't know why, but I knew I shouldn't hold her in front of all the rest of them.

Her pain was so loud that it took the fight out of everyone else, and they stood there awkwardly, not knowing what to do. Not wanting to see someone hurt like that, on display.

The DJ must've had an ounce of decency, because all of a sudden, the spotlight went out and the smoke machines blew a roiling gust of fog out onto the floor to hide us. I took my cue and headed for the door. There was nothing I could do to make it better, and the girls would no doubt be coming to put their arms around her, the one whose love hadn't yet turned to hate.

I pushed past Josh to get to the door, kicked it open—
and stopped.

There was a crowd out there on the sidewalk, and every last one of them turned to look at me. All of them men.

And I knew.

They were the husbands, the boyfriends, the brothers.

I saw in their eyes that they'd been waiting for me, and I knew they had been out there talking about what they were all looking forward to, and now that the moment had arrived, it caught them off guard. It was like the start of a lynching when every man knows what he is there to do, and he's waiting for someone else to make the first move.

I don't know why I didn't feel afraid.

To this day, I don't know.

There was only one way for me to go, and it wasn't back through that door. And I damn sure wasn't going to let a man touch me, no matter who he was. And I knew that if they had their chance, they'd destroy my face. They wouldn't be satisfied just kicking the shit out of me. They would make sure I never got kissed again.

And remember what I said about foreshadowing?

I pulled the baton out of my pocket, whipped it open into three feet of steel with a hard *snick!* and leaped at the closest one. I hit him hard across the side of his head in a downward stroke, seeing a dark line fill with blood, and struck another one in the teeth with a backhand. Pivoting on my heel and ducking a beer bottle, I punched someone else in the gut, rose, and bashed his nose into a bloody pulp with my forehead.

Someone grabbed the baton and I knew I couldn't waste time struggling to get it loose, so I jumping-snap-kicked some blonde guy in the face, turned, and drove my heel into the side of someone else's knee, hearing a snap.

With a desperate snarl, I managed to wrench my baton out of whoever's hands, and lunged to jab him in the eye with the knob on the end of it.

Reaching out blindly to grab someone else by the hair, I

swung with all my might and stove in at least one rib, then cast him aside like a rag doll. I was drunk with bloodlust. I'd never felt anything like it before. I knew I was going to die that night and was determined to take as many with me as I could. I sold my life as high as I might. I laid into them blindly and cracked bones with every stroke.

At some point I slipped on spilt booze and blood on the sidewalk and lost my balance. Thinking I had gone and done it now. I fought to keep upright, doing a dance that made all of my killers suddenly bold.

And that is when the first one crashed into me.

I guess the first sign of me losing gave the others a sudden courage, and those that had been holding back now came on in a rush. They hurled themselves at me, bearing me backwards to the side of the building, and dragging me down onto the sidewalk. More dog-piled onto me, just leaping and belly-flopping on top to crush us, but I managed to land on my side and keep a little breathing-space.

I snaked my hand into my other pocket and dragged out the bear spray. Fumbling with the safety, I pointed it as away from myself as I could in that tiny space, scrunched my eyes shut and held my breath, and squoze the button.

I felt it hiss all over my hand, felt it burn like my cheek was burning, and heard someone gasp.

"Omigod, get off me, get off me, get off me!" he cried.

But more people were dog-piling on top of us, the ones from the back of the crowd who wanted to be included before the fun was over. I needed air, but didn't dare take a breath. If I'd known, I don't know if I'd've sprayed underneath all those people, because holding your breath when you're proving how cool you are as a kid is a *lot* different than when you're panting from fighting twenty guys. There's no Zen-like tranquility of I-can-do-this. There's only Oh-shit-I'm-gonna-die.

My lungs were screaming for air, and I dimly heard what sounded like Josh shouting, from across vast gulfs of space.

I felt the crushing weight lighten a little, bit by bit, and I found out later that Josh had come outside and saved me. This part of it, apparently, had not been his idea and was something that happened on MySpace while all of the girls were planning the party. One had mentioned something to her boyfriend and he took the initiative of contacting every other guy until they had a lynch mob.

And Josh had walked outside to gloat, or whatever, just in time. I suppose I should be grateful for him to not've hated me so much. For him and the manager to've beaten up all the guys until they were off of me and running across the street to O'Shea's. But I'm not.

I remember getting up with a long, slow groan and dusting myself off, ignoring the hand someone had offered to help me up, and I staggered off to my car.

IX

"So, how was your Going Away party?" Elizabeth asked on the phone the next day. I guess she'd seen it on MySpace.

"Meh. It was all right," I said.

"Just *meh?*"

"Well, it was bittersweet."

"Oh, okay, I gotcha," she said, and dropped the subject. She read off a list of things she needed me to bring her, like a few tubes of Tom's of Maine, some Burt's Bees, a medium-sized dildo (I didn't ask why, thinking it was a test) and some wine-red queen-sized sheets. We chatted a little and finished the call off with some I-can't-waits and How-I've-missed yous.

When I hung up the phone, I took a deep breath, closed my eyes, and prayed.

"God," I said. "This time, I'm going to do it right. I will be the most honorable man you ever saw. I will never stray from this wonderful woman. I will never treat her unkindly or cheat on her or anything. I swear on my life that I will be pure and faithful, and that while I'm in Antigua, I will help the poor and do good things and go to Church and be honest.

"This curse they put on me last night is a joke. I have a good woman, the perfect woman for me, and we'll live happily ever after. And nothing will ever change that. Amen."

I opened my eyes and saw my Mom watching me from the red couch in the sitting room, a grey line of smoke curling up from the ashtray and wavering next to her eyes.

She hadn't said much to me lately, and now she had this look of accusation in her eyes. I felt a momentary panic that she had heard about last night, from someone, somehow. Felt the claustrophobic fear that the police were coming for me, as if I'd been the bad guy in the fight outside.

I swallowed hard, and managed a weak "Hi, Mom."

"So, you're leaving soon?" she asked.

I nodded.

"Just like that."

"Well, it hasn't been *just like that*. We've planned this for a good two weeks." Maybe I came off as a little harsh, because my nerves were fraying at the ends.

"You haven't planned it with *me.*"

"What? What does that mean?"

She heaved a world-weary sigh and waved dismissively. "Nevermind. Fine. *Fine.* You run along and do whatever you're going to do. Leave me here."

"Wait. Mom."

She waved more emphatically, making a face that was all drama-class resolve and dignity.

"It's time for you to leave the nest…"

Oh, is *that* what this is about? "Ma, I'm thirty years old."

"That…has nothing to do with it…"

And that's when I realized she was drunk.

At ten in the morning. No, wait, she wasn't up early and drunk already. She was up so late it was the next day and she hadn't gone to bed yet. She was so drunk she didn't notice my face was swollen and my eye was black. I must've come home and not noticed her lurking wherever she was. I just staggered in and went straight to my room.

She must've been listening to me packing yesterday, and that'd set her on this My-boy-is-leaving-me-*again* spiral. I knew she wanted me to go put my arms around her and tell her it was going to be all right, or something or other, but I couldn't find it in me. The fact that she wasn't happy for me, that she'd only thought of how it would affect her, made me cold.

That's right, I thought, maybe a little cruelly. No more "Oh good, you're home, you can go get me more cigarettes."

No more "I know you just got off work at one in the morning after a fifteen-hour shift, and have to be back at eight —" back when I had a restaurant job, "But you need to sit and listen to me talk about your father until the sun comes up."

And this was the biggest one, I guess, the loudest one:

No more having to lie about how I-don't-live-with-my-

Mother-my-Mother-lives-with-*me*.

Talk about emasculating. All that crap about me wanting to learn how to be a man? Goes out the window when you are saying "Shhh, baby, you'll wake my Mom."

Disgusted as much with myself as I was with her, I went up the stairs to my room and took another aspirin.

I drove around West Palm for what I was sure would be the last time, feeling dead to it all, and wondering every time I heard a police siren if it was meant for me. I bought all of the things Elizabeth wanted, feeling like a schmuck while getting the dildo, and didn't bother to stop and see any of my friends.

I figured, if they wanted to see me off, they would have thrown me a real Bon Voyage party. And *not* beaten me up.

I did go around and see my family though. I stopped off at my grandmother's, where my Dad and the Potato still were staying. Apparently, they were tweaking her will, since she was not looking very well. As soon as she got better, they'd be off to Sardinia again, where they were thinking of buying a house.

They wished me well, and I left. Then, my brothers, and my cousins, and then to the beach. I used to go there when I'd have something to think about.

The slow, lazy rhythm of waves crashing on the white sand always helped me put things into perspective. Something about the way they kept coming, all day and all night, without fail, reminded me that the world went on with its business and took no notice of me and my drama.

I stood there in the sand and the salty wind and curled my fingers for I don't know how long. That's something else I do, and something they don't tell you about in stories about us pickup artists. They glorify the wanton sex, but they don't tell you about the ghosts that haunt you, the girls you'll never hold again. If you see me staring off into space and making a claw of my fingers, that's me holding a hand that isn't there.

Oh boo-hoo, you might think. And you're probably right.

And maybe I am the only one who cares, who misses the

girls themselves and not just the conquests. But I have to admit, it doesn't change the fact that no matter which one I end up with, I'll always want the next one more. I will always be all starry-eyed and fascinated by that one over there.

I tried not to think about that on the plane, after turning off my phone. I had just hung up. Jen called to say good-bye. I won't go into any detail about our conversation. It's just one of those things you don't share. But I will say it was hard, and I know I will love her forever, in my way.

Looking out the window at the tarmac and the line of trees beyond it, and the wisps of cloud, I tried to think that it would never happen again. I would be faithful. I would be true to Elizabeth no matter what happened.

I suddenly decided to turn my phone back on to call my Mom and tell her she meant the world to me. Leave her behind on a happier note. It took forever, like it always does. Or maybe it took less than a minute, but it was the longest minute when you're waiting for something to boot up.

When it finally came on, before I could call her, it rang.

Startled, I answered it, wondering stupidly if it would be Tamara's mother, or the police.

"Hello?" I asked.

"Johnny?" It was Elizabeth. I sagged in relief, not realizing how much I had tensed up.

"Hi, sweetie."

"So, you're at the airport?"

"I am on the plane, waiting for takeoff."

"Good. I'll pick you up in *mas o menos* three hours."

"You'll be there?" I don't know why that surprised me. I guess because she took the train from the airport to visit me.

"Of course. So, I'll see you outside. There'll be a mob of people behind the railing so it might not be easy to see me but just so you know…"

"Okay. Thanks, honey."

"Aw, listen to you with the sweetie and the honey."

"Well…"

"So, I'm going to let you go, okay? They probably want you to turn off your phone."

"Okay. I'll see you soon."

There was silence on the other end for a moment.

"Hello?" I asked.

"Just wondering what I should say."

"What do you mean?"

"Well, this is too early for me to say 'I love you,' but you are doing this, moving to a different country to live with me."

"So...?" My mouth was suddenly dry.

"So, I'm not sure what to say."

"Say whatever you want to say."

"I'll see ya when you get here, champ. Safe flight."

I called my Mom and apologized to her, and she told me it was fine, and she'd be there if I needed her for anything. I thought about how unfair I had been to her, and said something along those lines. She had been the one whose credit card I paid for all the groceries with, every time I went. She had been the one who made sure the rent and light and water were paid every month. Yeah, I went to go get her cigarettes at three in the morning, but she carried me for nine months and stayed with my asshole Dad for so many years for my sake. Well, my brothers' too, but.

She allowed herself to become a hollow shell because of us. Some gratitude was in order. She said Don't mention it.

Some girl had called and left a message for me after I had left, though. I groaned and shifted in my seat.

"Yeah, hold on," Mom said. I heard papers rustling. "A moment after you left, she called. I figured I'd email you."

"Was it Jen?"

"No."

"Was it—"

"Keep your shirt on."

It was amazing that she could have lost a piece of paper she had just written on, while sitting at one table for hours. A grateful son would not think that about her, though, I reminded myself. So I took a deep breath and kept my mouth shut.

"Here it is," she said. "Tamara. Tamara called and said it was mean of her to play that little joke on you."

The bottom fell out from under me.

"She said she's fine, and living in Arizona now with her husband, and it was by sheer coincidence that she was visiting her parents and picked up the phone when you rang."

I couldn't speak.

I had to swallow, but my throat wouldn't work.

"So, there you have it," my Mom said. "I have done my only job for the day, delivered you this message, and I can take a break. Have a safe flight. Take lots of pictures. Let me know you got there safe, and all that."

I nodded dumbly, hanging up the phone, and looked at the tarmac out the window. Shaking my head slowly and starting to smile.

The trip was short enough, but I was left to wonder the whole time what Elizabeth meant. I agonized over it, debating all of the possible meanings to try and see where I stood with her, as if things had suddenly changed between us. When we arrived, I stood in several really long lines and struggled with my meager Spanish, and it would be as tedious in the retelling as it was in the experience, so I'll leave it out.

Then, customs. It seems that I stick out in a crowd, and my bags must be checked more thoroughly than anyone else's. Sure enough, three fat women in uniform called me over with great big smiles and asked me to open my luggage.

"Now, before you look in there," I said quietly, arousing their immediate suspicion. "There's a few things that I bought for my wife. All I ask is that you don't, you know, pick them up and wave them around for everybody to see."

They frowned and opened the first bag, and then their eyes lit up. There was that dildo on top, front and center, just like Elizabeth had instructed me, along with my other sex toys.

They started laughing and picking them out of the bag.

"Yeah, she's a wild one," I said sheepishly.

They started pretending that they were going to call out to someone and show what they had in their hands, and I put on a look of panic and shame, whispering "No no no no!"

They teased me a bit more, then did a perfunctory look under my clothes, saw my extendable baton and bear spray.

With their attention on the toys, they didn't even notice the weapons or check the other bags.

They grinned wickedly, zipped up my suitcase, and said "Welcome to Guatemala, loverboy!"

I smiled shyly at them and made myself blush a little.

Thinking So *that's* why she wanted me to bring a dildo.

Clever.

At the exit, I searched all the faces in the crowd of waiting families for her, trying not to look too eager.

And there she was.

Waitaminute.

That couldn't've been her.

Where were the pewter heels and the dove grey skirt and the wine red top? Where was the girl who stood up straight? I wouldn't have expected her to dress exactly like she had going out to dinner at Taboo, but I definitely expected her to dress better than this.

She had on dirty, torn jeans—and I don't mean torn *a la* the Ramones, I mean torn as in cheap jeans that'd been worn a lot more than they should've, through harrowing adventures, and she didn't have a better pair—and flip-flops and a ratty shirt, and some kind of hat that almost looked like a Roaring 20s flapper cloche, but was too cheap and dirty. She was pale, not like before, when she was Unblemished-by-the-sun, but I mean *pale,* like an albino. She had dark circles under her eyes, and she scratched at her nose, stretching her face and making an O with her mouth to open her nostrils wider. Scratching at the inside of her nostril. No, picking her nose. In front of God and everybody.

She noticed me looking and sniffed, making eye contact with me, but without any depth. Then gestured to the right. I looked and saw that's where the other people were headed, so I fell into the shoal of people.

When I got to her, she had a large and ugly dark man at her side, who sized me up with rapist's eyes.

"John, my *friend,* how was your flight?" she asked, coming in to give me a cold and lifeless kiss on the cheek. No, not an actual kiss. She just put her face alongside mine and made a sucking sound with her pursed lips.

What the f—?

Backing up to a respectable distance, she motioned with her hand to the ugly man and said "This is Toto."

The man I was supposed to avoid at all costs.

Right here at the goddamn *airport* the second I arrive.

And Elizabeth looked like hell. She slouched and stank, and was *not* the bright young woman I fell in love with.

Toto helped me with my bags and led us back to their car, where I squoze into the backseat next to a bunch of crap I guess had accumulated there over several years. She lit three of her cigarettes and passed two of them to Toto and me. He got his first, which annoyed me. He sat in the driver's seat, and it was unpleasant when he pulled out and whipped around all of the corners to get onto the road. It seemed he was deliberately driving like an asshole to make me nauseous.

The views of Guatemala on the road from the airport, I would find out later, are not the best the country has to offer, but when you've just arrived, it is a little, well, disconcerting. It makes you think you have arrived in a dump, and when you're met by a pair like this, it makes that crappy airplane food you ate turn sour in your belly.

They were talking, the two of them, but not to me.

I looked out the window at the spray-paint gangsta tags on corrugated zinc, held up as makeshift walls for cinderblock tumbledown huts. I looked out the window at dirty, grubby people, and, during the occasional break in trees, a primordial landscape with a smoking volcano in the distance.

And I couldn't help but think, what the hell did I get myself into?

Epilogue

Now, yunno when you're drinking maybe a little more than you should, and you time-travel? I think I said something about this before. You have even so much as one more sip of a drink, and next thing you know, you come to and you find yourself out on a cobblestone street a little ways down from a brightly-lit yellow arch with a clock in it. I know a lot of people have come out of their fast-forward on this street, because this is the street in Antigua with all the best bars on it.

I had no idea how I got there, where I'd been, or who I'd been with. Just *boom!* and there I was on Fifth Avenue. Oh, I'm sorry, Quinta Avenida. And my phone was ringing.

I don't know why, but I couldn't get it out of my pocket with my right hand, the side of my jeans the pocket is on, so I had to awkwardly dig for it with my left. I managed to work it out and answer it before it went to the voicemail I haven't yet figured out how to check.

The screen said it was Mali. I smiled and answered.

"Yello."

"Johnny! Where are you?" she yelled.

"Whoa! You about busted my eardrum!"

"You have to get off the street *now!*"

"What? Why?"

"The police are looking for you! They're saying that you killed somebody!"

"What? That's ridiculous!"

"Of course it is. But you know Toto has many friends in the police, and they are out for blood. You have to get out of sight and lie low for a while til we figure out what's going on. I can keep you at my place."

"Baby, this is insane."

"Welcome to Guatemala. Where are you now?"

I looked back at the arch and said "Fifth and Second."

I noticed people staring at me. Staring and pointing.

"Get off that street quickly and come to my place. I'll meet you there."

Click.

That girl was amazing. Best thing that ever happened to me, I thought.

I looked up at the people staring at me.

"What?" I shouted.

They bolted down the street the other way, tripping on the cobbles, and I watched them go, wondering what the hell their problem was.

Then I felt something heavy in my right hand.

Looked down and saw I was holding a gun.

Look for the next novel in the trilogy

MEMOIRS OF A SWINE

Just Plain Trouble

in paperback

It started innocently enough. Okay, no it didn't.

What happened, I was on my way to have lunch with Mali, the hot Norwegian chick with the faux-hawk, and I was halfway to the restaurant when I saw this little old Mayan lady, maybe four feet tall soaking wet, all bent and stooped over and bowlegged, struggling to carry two big baskets full of what must've been bricks for all the trouble she was having with them. She kept stopping to readjust her burden and stagger a few more steps.

I looked at her for a moment, thinking I really ought to help her, but I'll be late and…and then it hit me. How great it would be if I was late because I just had to stop and help some little old lady with her baskets. What a saint I'd seem to Mali!

I immediately felt terrible about it, thought about what I self-centered prick I was, and started to walk again. Keeping one eye on the little old lady. She was hobbling down the long colonnade on the south side of central park, and all I had to do was take two more steps for her to be out of my sight, and out of my life forever.

Then, hating myself for even considering ignoring her, I ran down the colonnade to help the unsteady woman. And again, for the record, I mostly did it because it was the right thing to do.

"*Señora!*" I said when I got to her side. "*Un momento.*"

She looked up at me with a very clear and cynical expression on her wizened face. I held out my hands and, not knowing how to say it in Spanish, asked her in English where she thought she was going with my baskets.

"*Como?*"

I thought for a second, then shrugged and just took the baskets from her, asking "*Donde vamos?*"

"Ahhh," she said, without wasting any time acting surprised or grateful. "*Gracias.*"

She waved me forward and started hobbling on, and I followed her. She tried a few times to engage me in conversation, but even if my Spanish was good I'd never be able to follow her dialect.

"*Lo siento, no hablo mucho español,*" I told her. I was getting pretty good at saying that, what with all the practice. She nodded, like she'd expected as much.

"*No stoy sorprendida, canche.*" Then she garbled a bunch of gibberish at me to pass the time until we reached the end of the colonnade. We had to wait a moment to let a horse and carriage go slowly by. The driver was a young Mayan fella who stared at the two of us as he passed, then nodded and gave me a thumbs-up.

Maybe this means I've watched too many movies, but I thought for a second that one day there would be a robbery, a bunch of guys with masks and guns, and one of them would be threatening me and demanding I hand over my money or he'd shoot me in the face, and this one robber would recognize me and go "Hey, not him. Let that one go." In Spanish, of course. And I would look at him and I'd recognize his eyes looking at me over the bandana he had on, covering the rest of his face.

Or maybe I'd be in a fight with Toto and I'd get a good lick in that makes the sonofabitch realize he's going to lose, and he pulls one of his guns out of thin air and sticks it in my face, and this guy would just happen to be there with twelve of his buddies. They'd all pull sawed-off shotguns out of their… wherever little Mayan dudes hide their sawed-off shotguns.

Then I thought, yeah, I definitely seen too many movies.

After he passed, there were a few cars that were spaced far enough apart that we could have crossed, but she seemed to be a little cautious. I felt bad again for being impatient, because she was still bent over and bowlegged and probably had good reason for not darting across the street.

When we finally made it across the cobblestones, back onto the sidewalk, damnedest thing, there was a really cute girl standing there watching us. Ceci, the suicide-blonde Antigueña who worked with Mali at Reilly's. With a big ol' grin splitting her face and showing me all of her perfect white teeth.

"So what're you, a...?" she asked when we got to her. "How do you say it? Good...Samaritan?"

I smiled and shrugged. "I have my moments."

"Ah, okay. You knock yourself out."

Perfect, I thought while we moved on at a snail's pace. She works with Mali, sees her every day, and she'll tell her all about me being such a nice guy. I couldn't've *paid* for better publicity than that.

Asshole, I thought immediately after. Don't do this because someone might see you. You do it because it's the right thing to do. But the selfish little voice in my head kept suggesting my karma would improve, if there was such a thing, or God would smile down upon me (again, if) and that I could go ahead and count my chickens, all the blessings that would rain down upon me for doing this one act of kindness.

And God *damn*, we were going a long way. Christ!

I looked back at the colonnade several blocks back and couldn't believe how long it had taken us to get that far, and wondered how much farther we'd have to go.

The baskets were getting heavy, and I scolded myself for thinking that, because I was one *hell* of a lot stronger than this little old lady, and she was going to make the trip all on her own without bitching about it. Probably made it by herself every day since forever. And would tomorrow, and every day until she died, carry these baskets full of bricks (it seemed) back and forth across Antigua.

I wondered what the hell she thought about that. Had she even bothered to dream about becoming a princess when she was younger, or did she already know she was going to carry heavy shit all over creation for her entire life? And was she resigned to it, or did she never even think there was an option? And was there an option at all?

And where the fuck was her *house*?

Christ Almighty! We had walked all the way to First Avenue and a bit past it, over the bridge to the forest that started across the easternmost street and sloped up the side of the mountain. She finally turned and said *"Muchas gracias, joven."* Gave me a little bow and took the baskets from me. I asked her if she was sure, and she nodded. Turned away and hobbled toward the forest.

I watched her go, and looked at my watch. Shit! I turned and ran back the way I'd come. A couple of times I ran across a street without looking and had to jump when I heard the screech of tires, feeling metal graze the fabric of my jeans and knowing I'd just barely missed being crippled. But I kept running. Thinking, damn, I must like this girl if I care that much about not keeping her waiting.

I felt the wind messing up the hair I'd gone to all the trouble of making perfect. I felt the heat of the sun and the exertion making my inner thighs start to sweat, and knew I'd need another shower before I could be comfortable taking my jeans off in someone else's company. And then, because I was already unclean, I went ahead and busted some parkour moves as I ran, every time I had to dodge something. Or someone. Vaulting over this and that, making a fool of myself, until I heard someone shout my name and I skidded to a stop.

Mali was standing on the other side of the street, in her tank top and faux-hawk, with her big happy smile.

"In a hurry somewhere?"

"Yeah, I got this really hot date and…" All of a sudden, my breath gave out on me, and my wind left and I was soaked in sweat, having to lean on a cyclopean wall behind me while Mali laughed.

"Sounds like every cigarette you've ever had is coming back to haunt you."

I wheezed at her and shook my head, grinning at how dumb I must've looked.

"Well, you can relax, honey. I'm running late too, as you may have noticed."

I started to say something clever, but doubled over and had to put my hands on my knees instead. There was this voice I was hearing in my head, a guy with an Australian accent I'd heard somewhere before, saying "Ah gotta tell ya, Ah've looked bettah." And I had to smile, shaking my head at myself.

So that was the start of it.

The second time I saw the little old lady, I was on my way home from work (I had gotten a job since then, teaching English) and I was going in that direction anyway, so I just said *"Hola. Donde vamos?"* again and took the baskets. Some people smiled at me as we passed.

The third time, I was in a hurry to my work and saw her on the sidewalk up ahead. I thought to myself, Fuuuuuck, not now, and crossed the street so that I could hurry past without seeing her seeing me, or at least pretend I hadn't seen her if she noticed. Again, I was ashamed of myself, and turned back around. Cursing myself because I knew I'd be late, I walked back to where she was readjusting her burden and stuck out my hands. She looked up and almost smiled. Almost, but not quite. Just gave me a little nod and handed the baskets over.

That time we went the other way, all the way cross town once again, but this time from central park to the market. We passed some young women who applauded from the window of the restaurant they worked in.

The fourth time, no one smiled at me, clapped, gave me a thumbs-up, none of that. I was disappointed, and even more disappointed in myself that I felt that way, and decided never to go out of my way to help anyone again, especially that little old lady, because if I was just doing it for attention or karma or God's favor, then it wasn't worth anything and was probably bad for my soul in some way.

So, I never did it again, ever.

Until the next week, when I saw her again, and it occurred to me she didn't give a shit why I helped her. She was just glad that I did.

And I thought, as I was taking her baskets again, man, this woman just *lives* to make me late, doesn't she?

One of the nights I was in Reilly's, this pretty Guatemalan girl came up to me and said in the Queen's English, "I saw you the other day, out on the street. You were helping an old woman with her baskets. You're a good man. I'm Ana. What's your name?"

And that worked out pretty well for me.

It got me thinking. The process of picking up a girl begins a long time before the pickup line. First, be friendly to everyone you see, and always be ready with a warm smile. That way, you will have laid the foundation in everyone's mind that you're worth knowing.

Every time I saw somebody pushing a stalled car down the street, trying to start it, from that moment on, I dropped whatever I was doing and lent a hand. That crippled guy who's always trying to take photos for tourists by the fountain in the park? Edwin? I carry him across the street and help him up to Doña Luisa, a block and a half away. I buy chuchitos from the lady who sits on the corner at night, hoping someone will buy them. I buy all of them, and leave them next to whoever I see sleeping on the ground outside the Muni.

I admit that I keep an eye out for hot chicks while I'm doing it, hoping they'll see me, but I don't think the people I help give a damn why I do it. They're just happy I do it. So, I say, instead of practicing pickup lines and palmistry, make an effort to be a better man.

Maybe the girl you'll want to meet later may have been someone you were nice to before. Or maybe it will be a friend of hers who will vouch for you that you are a decent guy. And who knows? Maybe you will have become one.

Alexander Ferrar was born on a battlefield, and the first sounds he heard were the shots that killed his parents. He was reared by a cutthroat band of evil mercenaries, and he grew up hard, suckled as he was at the teats of war. He—okay, no, he wasn't.

He lives in La Antigua Guatemala where he owns a restaurant-art gallery and a very popular exotic ice cream company, and he has a beautiful wife.
His grass is plenty green.

He's also the author of the *Heresy* series and *Icarus* trilogy of crime fiction novels, the sword-and-sorcery comedy *Saga of the Beverage Men*, the alternative history adventure *The Prince of Foxes*, and art collection *Variety is the Spice*.

But the other story sounded better, didn't it?

Made in the USA
Middletown, DE
21 April 2022

64585404R00123